THE HIGHLY SENSITIVE BRAIN

THE HIGHLY SENSITIVE BRAIN

Research, Assessment, and Treatment of Sensory Processing Sensitivity

Edited by

BIANCA P. ACEVEDO

ELSEVIER

ACADEMIC PRESS
An imprint of Elsevier

Academic Press is an imprint of Elsevier
125 London Wall, London EC2Y 5AS, United Kingdom
525 B Street, Suite 1650, San Diego, CA 92101, United States
50 Hampshire Street, 5th Floor, Cambridge, MA 02139, United States
The Boulevard, Langford Lane, Kidlington, Oxford OX5 1GB, United Kingdom

Library of Congress Cataloging-in-Publication Data
A catalog record for this book is available from the Library of Congress

British Library Cataloguing-in-Publication Data
A catalogue record for this book is available from the British Library

ISBN: 978-0-12-818251-2

For information on all Academic Press publications
visit our website at https://www.elsevier.com/books-and-journals

Publisher: Nikki Levy
Acquisitions Editor: Natalie Farra
Editorial Project Manager: Mona Zahir
Production Project Manager: Swapna Srinivasan
Cover Designer: Alan Studholme

Typeset by SPi Global, India

Contents

Contributors *ix*
Acknowledgments *xi*
Introduction *xiii*

1. The basics of sensory processing sensitivity **1**

Bianca P. Acevedo

1 The science of sensory processing sensitivity 4
2 The history of sensory processing sensitivity 5
3 Is high sensitivity a disorder? 10
References 14

2. Assessment of sensory processing sensitivity across the lifespan **17**

Francesca Lionetti

1 SPS in the context of the environmental sensitivity metaframework 18
2 Measures for the assessment of SPS in childhood and adolescence 22
3 Measures for the assessment of SPS in adulthood 30
4 Dandelions, Tulips, and Orchids: A metaphor of individual differences in sensitivity 36
5 Future directions 41
References 44

3. Sensory processing sensitivity—For better or for worse? Theory, evidence, and societal implications **51**

Greven U. Greven, Judith R. Homberg

1 SPS in the context of theories on environmental sensitivity 52
2 Empirical evidence for SPS as a marker for environmental sensitivity 63
3 Implications and future directions 68
4 Conclusion 70
References 71

4. Health and social outcomes in highly sensitive persons **75**

Jadzia Jagiellowicz, Sheerin Zarinafsar, Bianca P. Acevedo

1 Theories of SPS 75
2 Health and social outcomes in children and adolescents 76

3 Health and social outcomes in adults 88
4 Conclusion 97
References 98

5. Etiology of sensory processing sensitivity: Neurobiology, genes, and evolution 109

Flavia V. Naumann, Bianca P. Acevedo, Jadzia Jagiellowicz, Corina U. Greven, Judith R. Homberg

1 Evolutionary basis for sensory processing sensitivity 109
2 Genes underlying SPS 110
3 The neural basis of SPS 118
4 Peripheral sensitivities 125
5 Summary and future perspectives 126
References 128

6. Clinical assessment of sensory processing sensitivity 135

Elaine N. Aron

1 Assessment of highly sensitive persons 137
2 Working with HSPs 142
3 Typical issues working HSP issues in a clinical context 148
4 Personal relationships, career, and the workplace 156
5 Conclusion 162
References 163

7. Clinical characteristics of misophonia and its relation to sensory processing sensitivity: A critical analysis 165

Dean McKay, Bianca P. Acevedo

1 Basic conceptualization of misophonia 166
2 Defining the boundaries of misophonia 167
3 Emotional reactions associated with misophonia—Disgust and anger 173
4 Association of misophonia with other disorders 174
5 Sensory processing sensitivity and misophonia 179
6 Interventions and treatment of misophonia 180
7 Conclusions and future directions 181
References 183

8. The future of sensory processing sensitivity on planet Earth and beyond **187**

Bianca P. Acevedo

1 The future of sensory processing sensitivity beyond planet Earth 188

References 195

Appendix: Misophonia severity index *197*

Index *201*

Contributors

Bianca P. Acevedo
Neuroscience Research Institute; Department of Psychological and Brain Sciences, University of California, Santa Barbara, CA, United States

Elaine N. Aron
Department of Psychology, Stony Brook University, Stony Brook, NY, United States

Corina U. Greven
Department of Cognitive Neuroscience, Donders Institute for Brain, Cognition and Behaviour, Radboud University Medical Center; Karakter Child and Adolescent Psychiatry University Center, Nijmegen, The Netherlands; Social, Genetic & Developmental Psychiatry Centre, Institute of Psychiatry, Psychology & Neuroscience, King's College London, London, United Kingdom

Judith R. Homberg
Department of Cognitive Neuroscience, Donders Institute for Brain, Cognition and Behaviour, Radboud University Medical Center, Nijmegen, The Netherlands

Jadzia Jagiellowicz
Stony Brook University, Stony Brook, NY, United States

Francesca Lionetti
Department of Neurosciences, Imaging and Clinical Sciences, G. d'Annunzio University of Chieti-Pescara, Chieti, Italy; Department of Biological and Experimental Psychology, Queen Mary University of London, London, United Kingdom

Dean McKay
Department of Psychology, Fordham University, Bronx, NY, United States

Flavia V. Naumann
Department of Cognitive Neuroscience, Donders Institute for Brain, Cognition and Behaviour, Radboud University Medical Center, Nijmegen, The Netherlands

Eric A. Storch
Baylor College of Medicine, Compulsive, Obsessive, and Anxiety Program (COAP) Lab, Fordham University, Bronx, NY, United States

Sheerin Zarinafsar
Department of Psychological and Brain Sciences, University of California, Santa Barbara, CA, United States

Acknowledgments

For Gian Paolo and all other highly sensitive individuals of the world. May they all thrive on Earth, and beyond.

Introduction

Most of us know or may meet someone who is highly sensitive. In fact, approximately one in four people (20%–30% of the population) has an innate biological trait known as sensory processing sensitivity (SPS) or environmental sensitivity (ES), which is associated with greater responsivity to environmental and social stimuli, both for better and for worse. Because their systems are more responsive to stimuli, highly sensitive persons (HSPs) process information more deeply, make connections across themes, feel deeply, and perceive the "sights and sounds" of the world in an amplified way. Thus, the highly sensitive tend to become easily aroused and overstimulated by what others may experience as "normal." Because of this, highly sensitive individuals tend to prefer solitude over crowds, and thrive in calm and peaceful environments. However, they do tend to suffer more than others when they are exposed to negative, harsh, or impoverished conditions. This may pose serious issues for the highly sensitive because, as we all know, life is not perfect.

Therefore, most individuals with the SPS trait show symptoms at some point in their lives. Perhaps you or someone you know is experiencing some challenge or ailment that may be due to or exacerbated by high SPS. Symptoms of SPS manifest in a variety of ways, ranging from depression, anxiety, fatigue, and burn-out, to digestive issues and low self-esteem. Children with high SPS also manifest symptoms that overlap with attention deficit disorder, sensory processing disorder, and Autism.

If there is no awareness of the SPS trait, this may lead to a lifetime of "misunderstanding" for the highly sensitive. First, not being aware of the trait, many may simply view it as a disorder. Second, HSPs, mistaking their trait for a disorder or developing complexes such as social phobia or low self-esteem, may never learn to live and thrive in the world as they are: highly sensitive beings. Third, many HSPs may be advised to take prescription medications for symptoms setting them on a life-long trajectory of "having" to take something to function and altering their highly sensitive systems (which may have adapted, healed, and recovered from life's challenges on their own). There is much that has been written about the body's natural healing process, but that is beyond the scope of this book. However, it's important to note herein because complementary and alternative healing practices may be beneficial for SPS, as it is not a disorder.

Perhaps recognizing this intuitively, many highly sensitive individuals do turn to alternative medicine for healing and wellness.

A major goal of "The Highly Sensitive Brain" is to explain the science of sensitivity. Because our brains are viewed as the "control center" for the rest of our physiology, we focus on the neural underpinnings of SPS. However, our sensory system expands beyond the brain to our inner chemistry, the nervous system, and the energy that surrounds us (or what some practitioners call one's "aura"). In the future, there may be new research on SPS that may expand into these areas. However, in the current volume, we focus on the science of sensitivity and practical topics such as the theory, assessment, and clinical relevance of SPS.

As such, the aim of this book is multifold. It may serve as a practical guide for assessment, which may be particularly useful to clinicians, parents, educators, physicians, and healthcare workers. It may also serve as a tool for researchers and educators, providing them with background, theory, empirical evidence, and SPS measures which may be useful in furthering the science of sensitivity and its dissemination. "The Highly Sensitive Brain," is unique in that it is a culmination of knowledge from some of the top thinkers, researchers, and practitioners around the globe. These individuals have contributed their knowledge and insights for this first edition manual. This volume casts a large net to describe the history, biology, and assessment of this trait for a wide audience and also for specialists in their respective fields.

Chapter 1, "The Basics of Sensory Processing Sensitivity," provides a history and overview of SPS, with a mild evolutionary spin. Chapter 2 describes the theoretical foundations of SPS with corresponding self-report measures for assessing SPS in children and adults, and provides their psychometric properties across cultures. Chapter 2 also discusses the existence of sensitivity groups, along with the supporting empirical evidence. Chapter 3 delves even further into theoretical perspectives and research on SPS, introducing the umbrella term, "Environmental Sensitivity." Chapter 3 synthesizes the various theoretical frameworks of SPS to provide a current perspective of the field. It also discusses intervention studies and provides some implications of SPS for general public health and for building a diversity-aware society.

Chapter 4 moves to more practical work, detailing research on the physical, psychological, and social outcomes of highly sensitive children and adults. Topics include parent–child relationships, peer relationships, romantic relationships, and life and work satisfaction. It concludes with suggestions from the relationship level, to public health, and policy level for improving conditions to help harness the strengths of highly sensitive individuals.

Chapter 5 turns to the biology of SPS discussing the mechanics of SPS, detailing its neural and genetic basis. Chapter 5 explains how the neural underpinnings of SPS give rise to its phenotype with physical, emotional, and cognitive manifestations. The chapter closes with a discussion of what is needed in future to further the science of the SPS.

Chapter 6 takes a more practical focus, discussing clinical aspects of high sensitivity. This chapter covers four topics: (1) the assessment of SPS (by using an individual case as an example); (2) issues that clinicians often face when working with highly sensitive clients; (3) the three most common problems reported by highly sensitive clients in therapy; and (4) common relationship and work issues that occur among highly sensitive individuals.

Somewhat related, Chapter 7 discusses a specific condition, misophonia, which may seem related to SPS but is not. Misophonia (also known as "miso") is a selective sound aversion where individuals experience strong reactions to specific sounds. Its etiology, associated emotional reactions, psychological conditions, and psychopathological features are discussed, as well as its relation to SPS.

Chapter 8, the closing chapter, provides a perspective for the future of high sensitivity (SPS) taking a look at planet Earth and beyond. It provides an overview of the challenges of space travel and off-world exploration, and suggests how SPS may be harnessed (or difficult) for space missions. It closes with a discussion of what a future society might look like, and the ideals it may engender so that all individuals, including the highly sensitive, may thrive.

Bianca P. Acevedo
Neuroscience Research Institute, University of California, Santa Barbara, Santa Barbara, CA, United States

CHAPTER 1

The basics of sensory processing sensitivity

Bianca P. Acevedo
Neuroscience Research Institute, University of California, Santa Barbara, CA, United States

Contents

1 The science of sensory processing sensitivity 4
2 The history of sensory processing sensitivity 5
2.1 Why does variation in sensory processing sensitivity exist? 7
2.2 What are the advantages of high sensitivity? 9
3 Is high sensitivity a disorder? 10
3.1 What can we say about sensitivity now? 13
References 14

"If there is an "invisible hand" at work it is that empathy matures and consciousness expands to fill the temporal and spatial boundaries set by the new energy regime. Empathy becomes the thread that weaves an increasingly differentiated and individualized population into an integrated social tapestry, allowing the social organism to function as a whole." "The Empathic Civilization" by Jeremy Rifkin.

Sensitivity is a normal part of life. There are days when even the strongest of us may feel overwhelmed by the pressures of life, the world, our children, our bosses, and our heartbreaks. On the positive side, even the crankiest of humans may sometimes be overtaken by spontaneous joy, delight in small pleasures, a favorite song, or the sight of a beloved's face. However, the degree and frequency to which any individual experiences an amplified sense of, and perhaps oversensitivity to, the world and others varies. Some of us may feel very sensitive to things every day. We may become strongly affected by others' moods, hear every note in a song, and become bothered or delighted by the slightest of things. For others, it may take a great deal of stress, an extreme situation, or a specific trigger to induce a strong reaction. This enhanced sensitivity to the environment is also known as sensory processing sensitivity (SPS), differential susceptibility, vantage sensitivity, high environmental sensitivity, and biological sensitivity to context (see Greven

The Highly Sensitive Brain
https://doi.org/10.1016/B978-0-12-818251-2.00001-1

et al., 2019 for review). In addition to greater responsiveness to the environment, SPS is also characterized by enhanced awareness, empathy, depth of processing, and self-reflection (Aron & Aron, 1997; Acevedo, Aron, & Aron, 2018; Acevedo, Aron, Pospos, & Jessen, 2018).

At some level we all know that people vary in their sensitivity. I remember years ago, when talking to a nutritionist because I was having some gastrointestinal issues, the implicit nature of SPS came across in a very subtle way. At the time, I was having trouble digesting many foods which, once upon a time, had never bothered me. Suddenly, I could not eat many foods without experiencing pain and bloating. After completing a long list of questions, the nutritionist's diagnosis was not irritable bowel syndrome, nor a wheat allergy, nor a gluten intolerance. She proceeded to tell me in not so many words that she thought I was "too sensitive." The example she gave went something like this, "Bianca, my dad is the type of person who can have his finger chopped off and he will continue to do his job as if nothing has happened. On the other hand, my mother will cry when she gets a papercut." Although I was relieved to learn that I didn't have a chronic disease or a serious issue, I was mildly disappointed because there is no remedy for high sensitivity. It took years of trying different things to get my digestive system back to homeostasis. Eventually, Ayurveda, the ancient Indian practice of nutrition, is what worked for me; as well as figuring out that wheat was the culprit of many of my digestive issues.

Indeed, we all have varying degrees of sensitivity that is subject to change depending on the big and small circumstances of our lives, our life stage, and the types of activities we engage in. For example, mind-body practices, such as meditation, may enhance one's awareness of and sensitivity to the environment. However, for some individuals, (approximately 20%–30%), high sensitivity *is* their baseline. It is not a disorder and there is no cure for it. You can't take a magic pill to make it go away. If you are having challenges at work, or with your health, or your romantic partner, it is more likely than not that your boss, doctor, and even your romantic partner may misunderstand you, misdiagnose you, not get you, or flat out not have the mental or emotional bandwidth to support you. They may simply think that you're over-reactive, dramatic, eccentric, difficult, neurotic, or anxious. However, the flip side is that when SPS expresses itself adaptively, highly sensitive individuals engage in deep thinking, make better decisions, ponder spiritual questions, and engage in meaningful work (Aron & Aron, 1997; Aron, Aron, & Jagiellowicz, 2012). Thus, many people may deem those with the trait to be creative, gifted, empathic, nurturing, mystical, and intuitive.

Many of us may not be aware that we have the trait or that there is a genetic, biological basis for our sensitivity. SPS may be especially hard to identify in those of us that are extraverts or high sensation seekers [roughly 30% of highly sensitive persons (HSPs)] (Lionetti et al., 2018). Others may have been shunned for their high sensitivity, especially men (Falkenstein, 2019), so they learned to repress, deny, or hide it. And even for those who have a robust self-esteem about their sensitivity, they may not make a big show of it, but instead suffer silently when the world feels overwhelming. Also, they don't boast, brag, or give themselves enough credit when their sensitivity results in great acts of kindness, creativity, insight, and love. Indeed, many highly sensitive individuals suffer from low self-esteem (Aron et al., 2010) as studies have shown significant correlations between high SPS and low self-esteem (Acevedo, Aron, & Aron, 2018; Acevedo, Aron, Pospos, et al., 2018). Some has been written about this topic, for example, see Dr. E Aron's book, *The Undervalued Self*. Also, stigma, low self-esteem, and shame are commonly experienced by highly sensitive individuals as a result of knowing, whether implicitly or explicitly, that they are "different," a minority, and being misunderstood.

The aim of this book is multifold. One of the aims of this volume is to shine a light on the science of sensitivity so that individuals with the trait and others can come to understand it as a normal variation that does not need fixing, shunning, or special accommodations. It is simply a variation, like being right-handed versus left-handed, or being tall versus short. These variations all serve a purpose and many have remained in our gene pool because they have helped with species survival in some way.

To achieve these aims, this book describes theory, research, validated assessment tools, and clinical issues related to high sensitivity. It may serve as a practical manual for the assessment of SPS which may be particularly useful to clinicians, educators, physicians, and health care workers. It may also serve as a tool for researchers and instructors, providing them with background, theory, and science on SPS to date, so they may apply the measures and ideas in their research and/or the classroom.

Another goal of this volume is to provide a comprehensive scientific analysis of SPS so that more individuals may learn to understand and have a greater respect for this trait, rather than seeing it as a weakness or deficiency. This may help future generations to simply "be" of their kind and realize that when those with the trait are nurtured, they have much more to offer, and the world to gain. However, it's important to realize that it is pretty standard and normal for humans to have preferences, both slight and strong.

Also, the problem with any minority is that humans have a hard time not falling into one group or another. It makes things, situations, and people easier to process cognitively when they are compartmentalized so that we can quickly size things (and people) up without taking the extra cognitive effort to unpack it all. Indeed, this is yet another strategy that has helped our species to survive.

1 The science of sensory processing sensitivity

According to scientists, there are a few big questions that remain to be answered, "What is the origin of life?", "What is the universe made of?", "Is there intelligent life beyond our planet", and "What is the source of human consciousness?". I would argue that the topic of this book indirectly addresses the last of these questions as it largely deals with a scientific explanation of SPS: an innate, genetically based trait that is characterized by enhanced awareness and responsivity to the environment and other individuals. Other cardinal features of SPS include heightened awareness, enhanced responsivity to the environment, and greater depth of processing, self-reflection, empathy, and intuition. Coincidentally, some of the characteristics attributed to SPS overlap with those that scientists consider to engender "mental consciousness," such as the ability to be aware that one is experiencing something. Mental consciousness is thought to be distinct from creature consciousness (or wakefulness) in that it involves awareness that a stimulus is present, awareness that one's self exists, and awareness that it is one's self that is sensing, behaving, and solving problems (Ledoux & Brown, 2017). Therefore, by gaining a better understanding of the biological mechanisms underlying SPS, we may inadvertently contribute to solving "the hard problem of consciousness."

In seminal research on the neural basis of SPS, it was discovered that as a function of the SPS trait, greater brain response was shown in brain regions associated with awareness, memory, self-other processing, and empathy (Acevedo et al., 2014; Acevedo, Jagiellowicz, Aron, Marhenke, & Aron, 2017; Jagiellowicz et al., 2011). Applying some of this knowledge to understanding the hard problem of consciousness may help scientists, for example, by providing them with a more narrow focus of brain regions that may be targeted for the study of mental consciousness. Also, perhaps by focusing our examinations on individuals, who are both highly sensitive and well adjusted, thanks to good-enough childhoods and not-too-stressful lives, we may gain a better understanding of the expression of the trait in its purest form. Indeed, in positive environments, high SPS persons can be quite content, calm, and well-adjusted. Not that I mean to say that highly sensitive people are always more calm or aware than others, but again, such work is needed. Another issue is that most of the world sees dysfunction more readily because unhappy individuals tend to complain more, and express their stress, discomfort, and anxiety. On the other hand, highly sensitive individuals tend to like their "solitude," so we may not see them at their best. Again, such work may further advance the scientific understanding of consciousness.

2 The history of sensory processing sensitivity

The road to a scientific understanding of SPS has not been linear. It has taken interesting twists and turns. In its infancy, the systematic study of what we now call SPS was largely conducted by researchers and psychologists studying temperament in children who coined the terms "inhibited," "introverted," and "over-responsive" to describe the trait. They observed that "over-responsive" children tended to respond to novel or stressful situations by retreating, crying, or approaching the new situation slowly and with caution. Researchers also observed that the so-called inhibited or highly responsive children were more affected by their environments, including family members, such as siblings, both for better and for worse. When the childhood environment was relatively stable, positive, and nurturing the children tended to adjust to the stressors, remain calm, and show normal signs of development. However, when exposed to harsh, chaotic, or unsupportive environments, the "inhibited" children tended to show even worse adjustment compared to those less inhibited (Biederman et al., 1990; Fox, Henderson, Marshall, & Nichols, 2005). For example, they tended to become reclusive and show abnormal or delayed development, including having poor coordination, delayed speech, and deficiencies in social behaviors.

In the 1990's research on high sensitivity took a different turn through the work of clinical psychologist Dr. Elaine Aron, a pioneer in the study of sensitivity in adults. At the time, Dr. E. Aron, being highly sensitive herself and working with several highly sensitive patients, realized that what many researchers and clinicians had described as "shyness," "introversion," or "anxiety" was in fact a lot more complicated. Thus, she conducted a series of in-depth, systematic interviews of people who recognized themselves as "highly sensitive persons." She realized a better term for this trait might be high sensitivity, and the interviews that led her to that conclusion resulted in the creation of The HSP scale, a reliable 27-item self-report measure of

SPS (Aron & Aron, 1997). This was a game changer for the scientific study of this trait, as it gave it a new name and theory behind it, and allowed quick assessment of the trait with a self-report scale. The HSP scale has been validated, is widely used now, and will be discussed in greater detail in Chapter 2 on the assessment of SPS.

Another addition to the field resulting from Dr. E. Aron's work was the coining of the more easily relatable term "the highly sensitive person." Her book and the HSP scale provided a strong basis for understanding positive and negative aspects of the trait, such as sensitivity to stimuli and others' emotions, depth of processing, self-reflection, conscientiousness, creativity, and an affinity for aesthetics; as well as the tendency to become overwhelmed when having too much to do, having a lot of incoming information, and having to compete or be observed. Thus, it made the information, which had an empirical basis, accessible to the public. Also, it characterized the trait as a variation in information processing, rather than a disorder; and distinguished it from other personality traits such as Neuroticism. Dr. E. Aron concluded that highly sensitive individuals were not fragmented, frazzled, overwhelmed, shy, or introverted; but rather that they could appear this way and manifest these symptoms in overly stimulating situations or if raised in environments that were overly stressful, harsh, abusive, or impoverished.

Thus, Dr. E. Aron's work on high sensitivity revolutionized the way many came to view themselves and others with the trait. It provided the basis for many to communicate with each other about the experiences they had been having all their lives, but had never put into words. Partly, this may be due to being a highly sensitive child without HSP parents or caregivers. Without a highly sensitive subculture around them, many children go without the language and scaffolding needed to understand their trait.

In parallel to Dr. E. Aron's work, researchers were also making observations of "high sensitivity" not only in humans, but also in over 100 other animal species (Wolf, van Doorn, & Weissing, 2008). The result was the birth of several theories to describe the trait. Chapters 2–4 will explain the theoretical models of differential susceptibility, biological sensitivity to context, SPS, and environmental sensitivity. These theories and supporting research suggest that "high sensitivity" is a normal variation in sensory processing with biological underpinnings that result in individual differences in responsivity to the environment and social stimuli. As such, these theories suggest that HSPs are more susceptible both to the detrimental effects of harsh environments, and the benefits of positive and nurturing ones. As such, early childhood environment plays a particularly important role in the lives of HSPs, having significant implications for lifetime outcomes.

Due to a higher degree of sensitivity, those with the trait generally observe subtleties in the environment and others (Jagiellowicz et al., 2011). They may perceive sights, sounds, smells, and others' emotional expressions more strongly (Aron et al., 2012; Jagiellowicz, Aron, & Aron, 2016). In addition to perceiving stimuli more strongly, highly sensitive individuals also process the incoming information more deeply (Acevedo et al., 2017; Jagiellowicz et al., 2011), such as making connections across events, gaining deep insights, having a deeply ingrained memory of the event, and seeing that which is beyond the mind's eye. With a deeper processing and understanding of environmental and social information, highly sensitive individuals are more affected by the environment, situations, and others—both for better and for worse (Acevedo et al., 2017; Aron et al., 2012; Jagiellowicz et al., 2016).

For example, highly sensitive individuals may derive great joy from aesthetically pleasing things, ranging from artwork to music, to a lover's touch, a beloved's smile, or the taste of a fine wine. In the brain, these experiences register as greater reward, emotion, and memory signaling (Acevedo et al., 2014, 2017; Acevedo, Aron, & Aron, 2018; Acevedo, Aron, Pospos, et al., 2018). However, due to the increased depth of processing and memory that these processes require, highly sensitive individuals may tire and become overwhelmed more easily than those less sensitive; and even in response to positive things such as a birthday party, a wedding celebration, or an intense erotic experience (Jagiellowicz et al., 2016). Thus, the trait does have inherent complications, such as requiring more downtime to integrate information, feeling tired or drained when experiences require a lot of processing, and the tendency to become overwhelmed in overstimulating situations. Also, highly sensitive individuals tend to experience heightened stress and reduced reward when exposed to negative things, such as the scene of an accident, a foul smell, or rejection; but also they show greater reward brain response to positive stimuli (Acevedo et al., 2017; Jagiellowicz et al., 2016).

2.1 Why does variation in sensory processing sensitivity exist?

Why variations in SPS evolved, and high sensitivity has continued to persist in humans and other species, given its notable drawbacks, has been a topic of increasing discussion in recent times. One idea that has been proposed by theorists and some scientists is that high sensitivity is a strategy that *may* be adaptive under some conditions (Aron et al., 2012). For example, the more sensitive organism, being more aware of toxic elements, threats, and harm, may avoid exposure to these elements thus increasing their fitness and survival. Also, by being more aware in general, high SPS persons may be more alert to opportunities, resources, and rewards; as well as having the memory for these events, such as where to find them or how to get them.

Also, high sensitivity may be adaptive as individuals with the trait are very empathic. Empathy facilitates stabilizing cooperative relationships and trust in humans and other species (e.g., McNamara & Houston, 2009). Indeed, cooperation is thought to be adaptive, especially toward kin and close others (Preston, Hofelich, & Stansfield, 2013; Raghanti et al., 2018; Wilson, 1975), as it grants benefits—including increased mating success, social standing, and the sharing of resources (Fehr & Rockenbach, 2004)—to the "giver" as well as the recipient. "Altruistic" traits are thought to prevail as the genes that the more altruistic individuals share with kin would be passed on (Nowak, 2006). However, the right balance of empathy is necessary to respond to others' needs without becoming overly distressed, something that high SPS persons tend to do if they have not developed appropriate coping strategies.

2.1.1 Why is high sensitivity a minority?

High sensitivity is not adapted by all organisms. Evidence suggests that high SPS appears in the minority within a given species, with estimates ranging between 20% and 30% (Aron & Aron, 1997; Aron et al., 2012; Belsky & Pluess, 2009; Kagan, 2002; Suomi, 1997; Wolf et al., 2008). Initially, research largely used a two-factor category system where individuals were placed in either a "highly sensitive" or "low sensitive" group. More recent work with humans suggests a three-part category system with somewhat different estimates, placing 30% of individuals as highly sensitive, 40% as moderately sensitive, and 30% as low on sensitivity (Lionetti et al., 2018; Pluess et al., 2017). Given that SPS appears to be somewhat beneficial for individuals and society, why is it a negative-frequency-dependent trait, meaning that it is only found in a minority of individuals?

One idea proposed by scientists for SPS being a negative-frequency-dependent trait is that if all organisms were equally sensitive, there would be no advantage to it. Another possibility is that variation in degrees of sensitivities is needed for the group to prosper under various conditions, both supportive and abundant, but also harsh and impoverished. Take, for example, a stressful or harsh environment with extreme temperatures, violence, and uncertainty. Under these circumstances, individuals with low sensitivity may have the advantage as they may be better equipped to carry on, explore, and survive. Even in milder, but still overstimulating circumstances, such as riding a crowded metro, attending a concert, or walking through a crowded museum, having a less sensitive companion to lead the way may be comforting. Thus, in environments that are less conducive for high sensitivity—groups, species, and individuals may benefit from having individuals with low sensitivity. In situations where highly sensitive individuals may not be able to respond to a situation due to being overwhelmed, and feeling emotionally and cognitively taxed, the less sensitive person or partner may pick up the load. Indeed, as will be discussed throughout this volume, high sensitivity has significant cognitive, emotional, and physiological costs associated with it.

Thus, all levels of sensitivity are needed as they have advantages that vary by the circumstance and ultimately all have the potential to benefit society. One strategy is not better than the other. Variation in sensitivity is needed, both strategies are equally valuable, and it's unlikely that one could survive without the other. Thus, neither side ought to expect the other to concede, change, or be dissolved. Instead it seems most beneficial when each side can give and also take the best from the other.

2.2 What are the advantages of high sensitivity?

Kin, close others (such as partners or friends), and group members may also benefit from having a highly sensitive friend, as the high SPS friend is more likely be attuned to and responsive to their needs (Acevedo et al., 2017; Aron et al., 2012). This has the clause, as will be discussed in forthcoming chapters, that the highly sensitive partner will be best equipped to respond appropriately if they have had a positive childhood and they are not overwhelmed. Under ideal conditions, highly sensitive individuals may respond to stressful situations, including others' distress, without becoming overwhelmed, burdened, or undone.

Beyond being empathic to other's emotions and needs, highly sensitive individuals enrich the environment that we all share in other ways. For example, being aware of subtleties in the environment (Aron et al., 2012; Jagiellowicz et al., 2011), they are more likely to make slight adjustments that may increase comfort for all, such as dimming the lights, moving the angle of a painting, or simply cleaning up so that the environment is experienced as more organized (and less chaotic). They are also more likely to recognize injustices, and even with their usually reserved, calm, and quiet demeanor, highly sensitive individuals are more likely to make bold moves to stand up for the weak and remove themselves from harm's way. Not surprisingly, the responsiveness, empathic, and action-oriented aspects of high sensitivity are observed in the brain (Acevedo et al., 2017), and some of these characteristics had gone undernoticed due to a large reliance on behavioral research examining clinical aspects related to high SPS.

In summary, high SPS, despite having some limitations, may have several advantages when the trait is expressed adaptively. For example, being more aware of subtleties in the environment (Aron et al., 2012; Jagiellowicz et al., 2011), highly sensitive individuals may be more perceptive of and remember how to obtain rewards, resources, and alliances. Also, they will be more aware of harm, threats and foes. Thus, they will have the advantage of knowing how to best respond to the environment, as well as being able to share this information with others. This, as well as their proclivity for empathic concern, will enhance cooperation. Thus, even when they are not able to fend for themselves due to exhaustion, physiological dysregulation and the like, the highly sensitive may have the advantage of having had formed alliances.

3 Is high sensitivity a disorder?

It's interesting to note that with the rise of analytical, rational, and masculine-dominant societies, that in modern times, the more gentle, intuitive, and sensitive aspects have come to be devalued and marginalized. Why some humans would be motivated to denigrate, oppress, or even obliterate these qualities, or individuals that express these characteristics, is a topic that requires a different forum for discussion. For example, a few centuries ago our fellow ancestors participated in "witch-hunts" where many women were persecuted and put to death with the accusation that they were engaging in dark magic. Interestingly, these individuals may have been highly sensitive as they were charged with having healing abilities, being able to see into the future (intuition), and having magical powers (spirituality). Perhaps the marginalization of HSPs stems from fear, lack of understanding, or feeling that in order to survive in harsh environments, sensitivity in any individual must be minimized. Thus, it is not surprising that in times of war, famine, and scarcity the more rational, unemotional, and cold side of humanity has prevailed.

However, in kinder times, research on humane themes such as love, altruism, and sensitivity has had the opportunity to flourish. Today, even with many global shortcomings and increasing disparities, people's health, well-being, and mental stability are monitored and addressed globally by organizations such as the United Nations and World Health Organization. However, there is a trend toward overdiagnosis of disorders, largely in societies where the pharmaceutical industry has a stronghold. Thus, it is not surprising that many people are diagnosed with "disorders" ranging from depression, anxiety, and chronic fatigue, to attention hyperactivity deficit disorder.

Although there are certainly times when prescription medications are needed to help individuals return to or maintain a state of homeostasis, a tendency toward overdiagnosis of mental health issues may set many whom could have recovered naturally, on a life-long trajectory of taking prescription medications for their mental health. This is especially pertinent to highly sensitive individuals that have had negative childhoods, as they do show higher incidences of depression, anxiety, and diagnosable disorders (see Chapter 4 of this volume for review). However, in some cases they may simply be responding to life events in a normal way for them—such as by feeling extreme sadness, loss, or disorientation to a life-changing event. After all, what is deemed normal is naturally going to be the way the majority of people react. However, the highly sensitive may need extra support or time to recover from stressful life events. Thus, lack of awareness about the trait does not always allow for the best healing practices for HSPs. But, many individuals do follow the road less traveled to alternative (largely non-Western) medical practices, and I believe that there many of you reading this book. That is not to say that doctors, teachers, or researchers that misdiagnose sensitivity for a disorder and overprescribe medications are mean spirited. They are simply not aware of the trait themselves. Temperament as a factor to be attended to is not taught in professional schools. Thus, practitioners often commingle SPS with anxiety, depression, ADHD, autism, and other seemingly related disorders.

For example, "sensory processing disorder" (SPD) is a condition where individuals' experience hypersensitivity to light and sounds, and may have a hard time integrating sensory signals. Unlike HSPs, those with SPD have difficulties communicating, socializing, and mobilizing. Treatment for SPD often involves physical, behavioral, or cognitive therapy and is largely focused on stabilizing sensory signals and mobility. Highly sensitive individuals do not show poor coordination or disorientation to sensory signals, but they might if they had SPD, although SDP is not necessary for SPS to occur. When overwhelmed, HSPs may feel so tired and overstimulated that they cannot adequately process more information. However, the baseline for high SPS persons is not dysregulation of sensory processing but rather deeper sensory processing. Thus, for SPS no specific treatment is offered.

Chapter 6 on clinical issues related to SPS will discuss how psychotherapists might approach their work with highly sensitive clients, with or without other symptoms. Namely, this involves an understanding of the trait, lifestyle modifications, and treatment of co-occurring disorders. Also, throughout the volume we will discuss how some cognitive/behavioral and complementary/alternative interventions may more fruitfully benefit the highly sensitive. Although these are not "treatments" per se, there is some solid preliminary evidence suggesting that cognitive and behavioral interventions may increase resiliency and buffer highly sensitive individuals from negative outcomes such as reduced cognitive performance, fatigue, burnout, anxiety and depression.

As will be discussed in depth in Chapter 4 on outcomes related to SPS, highly sensitive individuals commonly experience the symptoms and issues noted earlier (anxiety, depression, chronic fatigue, burnout, etc.). It is very likely that the higher prevalence of these disorders among the highly sensitive may be due to having had negative childhoods and developing negative coping strategies, as well as misdiagnosis. However, the trait itself is not the symptom, but rather these symptoms may emerge when the individual experiences continued exposure to negative, toxic, or stressful environments.

In recent times, coinciding with an increased incidence and awareness of autism spectrum disorders (ASD), especially in children—many high SPS persons also may have been mistaken for having ASD. Certainly, highly sensitive individuals can show symptoms related to autism, such as experiencing hyper-responsiveness to stimuli and high sensory sensations. However, one key way that SPS is distinct from ASD deals with empathy and response to social stimuli. Indeed, a signature feature of SPS is enhanced empathy and responsivity to social stimuli, including others' emotions and signals (Acevedo et al., 2014, 2017). Highly sensitive individuals are usually very aware of others' moods, emotional states, and subtle cues (Aron et al., 2012; Aron & Aron, 1997). They can perceive someone's sadness, anger, confusion, and joy quite readily and we also have neural evidence of this. In a series of studies, my collaborators and I showed that those with greater SPS showed stronger reward signals in response to others' happy face images, and also more activation in areas related to empathy in response to others' sad faces. Moreover, these neural signals were amplified both with increasing sensitivity and in response to a close other (versus a stranger), and also with self-reports of a more positive childhood. In stark contrast, a key feature of

ASD is a deficit in the ability to understand others' emotional and social signals. There is evidence of this in children with ASD, as well as neural evidence showing that the pathways that usually light up in response to social and emotional cues, are dysregulated in those with ASD. Thus, although both SPS and ASD show hyper-responsiveness in some situations, they are distinct in this one very important way: the response to social cues. It's especially important for parents, teachers, and physicians to be aware of this, so they do not mistakenly treat a highly sensitive child for ASD. Ideally, it is best to obtain an expert opinion from a professional that understands both ASD and temperament.

3.1 What can we say about sensitivity now?

In recent decades, research, theory, and public awareness regarding SPS have grown significantly. The rise in interest for SPS coincides with rapidly changing social norms whereby individual differences are more accepted, or at minimum, there is greater awareness that people vary in their preferences, tendencies, personalities, and temperaments. Also contributing to increased interest in SPS are the rapid global changes taking place with technological innovations, globalization, extreme climate changes, and a renewed emerging interest in human missions to the moon and other planets (Szocik, Wójtowicz, Boone Rappaport, & Corbally, 2020). However, it is important to balance these views with the enormous capacity that our species has displayed such as great acts of altruism, beauty, and wonder. We have mastered language, the arts, and also managed to make a luxury of self-development, self-expression, and awareness. We display great flexibility when it comes to love, and beyond mating and offspring have shown that coupling and sensitivity serve many and complicated purposes, which ultimately have contributed to the advancement and thriving of our species.

However, even with all the knowledge and reason we have garnered it would be unfair to say that we are not facing changes and challenges never experienced before. Predictive models and big data are useful, but they must be tempered with wisdom, good judgment, and morals. We are in interesting times where science and intuition meet, and it's imperative that decision makers use both to guide their thinking. It's interesting to note that with all the changes occurring, many individuals seem to be adapting with what appears to be an enhanced sensitivity. It is possible that high SPS, with its predisposition to responding and adapting to the environment could be beneficial when considering exploring new environments?

In recent years, there has been a rebirth (since the 50s) of considering human space travel and possible settlements on Mars and Luna (Earth's moon). Scientists discussing space exploration consider "human enhancements" to be critical for space travel, exploration, and colonization (Szocik et al., 2020). Human enhancements include, for example, the insertion of some genes into humans so that they may have a resistance to radiation, which is needed to survive Mars' radiation-dense atmosphere. Other enhancements may include prosthetic or genetically engineered capabilities designed for living in a specific space environment, with extreme dryness or microgravity. Thus, investigating the physiology and biological mechanisms underlying SPS—with enhanced awareness, intuition, reflective thinking, and a general contentment with solitude—may be beneficial in considering human enhancements and qualities that may be suitable for spaceflight and off-world settlement. Still, we have much work to do here in terms of exploring SPS on planet Earth. Certainly, this is a good starting point for this volume, and for considering the possibilities of sensitivity here on Earth and beyond.

References

Acevedo, B., Aron, E., Aron, A., Sangster, M., Collins, N., & Brown, L. (2014). The highly sensitive brain: An fMRI study of sensory processing sensitivity. *Brain and Behavior, 4*, 580–594.

Acevedo, B., Aron, E., Pospos, S., & Jessen, D. (2018). The functional highly sensitive brain: a review of the brain circuits underlying sensory processing sensitivity and seemingly related disorders. *Philosophical Transactions of the Royal Society London B Biological Sciences, 373*(1744).

Acevedo, B. P., Aron, E. N., & Aron, A. (2018). *Novel perspectives on Sensory Processing Sensitivity*. San Francisco, CA, USA: Association for Psychological Science Convention.

Acevedo, B. P., Jagiellowicz, J., Aron, E., Marhenke, R., & Aron, A. (2017). Sensory processing sensitivity and childhood quality's effects on neural responses to emotional stimuli. *Clinical Neuropsychiatry, 14*(6), 359–373.

Aron, A., Ketay, S., Hedden, T., Aron, E. N., Markus, H., & Gabrieli, J. E. (2010). Temperament trait of sensory processing sensitivity moderates cultural differences in neural response. *Social Cognitive and Affective Neuroscience, 5*, 219–226.

Aron, E. N., & Aron, A. (1997). Sensory-processing sensitivity and its relation to introversion and emotionality. *Journal of Personality and Social Psychology, 73*(2), 345–368.

Aron, E. N., Aron, A., & Jagiellowicz, J. (2012). Sensory processing sensitivity: a review in the light of the evolution of biological responsivity. *Personality and Social Psychology Review, 16*(3), 262–282.

Belsky, J., & Pluess, M. (2009). Beyond diathesis stress: Differential susceptibility to environmental influences. *Psychological Bulletin, 135*(6), 885–908.

Biederman, J., Rosenbaum, J. F., Hirshfeld, D. R., Faraone, S.V., Bolduc, E. A., & Gersten, M. (1990). Psychiatric correlates of behavioral inhibition in young children of parents with and without psychiatric disorders. *Archives of General Psychiatry, 47*, 21–26.

Falkenstein. (2019). *The highly sensitive man: Finding strength in sensitivity*. New York: Citadel Press Books.

Fehr, E., & Rockenbach, B. (2004). Human altruism: Economic, neural, and evolutionary perspectives. *Current Opinion in Neurobiology, 14*, 784–790.

Fox, N. A., Henderson, H. A., Marshall, P. J., Nichols, K. E., & Ghera, M. M. (2005). Behavioral inhibition: Linking biology and behavior within a developmental framework. *Annual Review of Psychology, 56*, 235–262.

Greven, C. U., Lionetti, F., Booth, C., Aron, E., Fox, E., Schendan, H. E., ... Homberg, J. (2019). Sensory processing sensitivity in the context of environmental sensitivity: A critical review and development of research agenda. *Neuroscience and Biobehavioral Reviews, 98*, 287–305.

Jagiellowicz, J., Aron, A., & Aron, E. N. (2016). Relationship between the temperament trait of sensory processing sensitivity and emotional reactivity. *Social Behavior and Personality: An International Journal, 44*(2), 185–199.

Jagiellowicz, J., Xu, X., Aron, A., Aron, E. N., Cao, G., Feng, T., & Weng, X. (2011). The trait of sensory processing sensitivity and neural responses to changes in visual scenes. *Social Cognitive and Affective Neuroscience, 6*(1), 38–47.

Kagan, J. (2002). Childhood predictors of state anxiety. *Dialogues in Clinical Neuroscience, 4*(3), 287–293.

Ledoux, J., & Brown, R. (2017). A higher-order theory of emotional consciousness. *Proceedings of the National Academy of Sciences of the United States of America, 114*(10), 2016–2025.

Lionetti, F., Aron, A., Aron, E. N., Burns, L., Jagiellowicz, J., & Pluess, M. (2018). Dandelions, tulips and orchids: evidence for the existence of low-sensitive, medium-sensitive and high-sensitive individuals. *Translational Psychology, 8*(1), 24.

McNamara, J. M., & Houston, A. I. (2009). Integrating function and mechanism. *Trends in Ecology & Evolution, 24*, 670–675.

Nowak, M. (2006). Five rules for the evolution of cooperation. *Science, 314*, 1560–1563.

Pluess, M., Assary, E., Lionetti, F., Lester, K. J., Krapohl, E., Aron, E. N., & Aron, A. (2017). Environmental sensitivity in children: Development of the Highly Sensitive Child Scale and identification of sensitivity groups. *Developmental Psychology, 54*, 51–70.

Preston, S. D., Hofelich, A. J., & Stansfield, R. B. (2013). The ethology of empathy: A taxonomy of real-world targets of need and their effect on observers. *Frontiers in Human Neuroscience*, 7, 488.

Raghanti, M. A., Elder, M. K., Stephenson, A. R., Munger, E. L., Jacobs, B., Hof, P. R., ... Lovejoy, C. O. (2018). A neurochemical hypothesis for the origin of hominids. *PNAS, 115*(6), 1108–1116.

Suomi, S. J. (1997). Early determinants of behaviour: Evidence from primate studies. *British Medical Bulletin, 53*(1), 170–184.

Szocik, K., Wójtowicz, T., Boone Rappaport, M., & Corbally, C. (2020). Ethical issues of human enhancements for space missions to Mars and beyond. *Futures, 115*, 1–14.

Wilson, E. O. (1975). *Sociobiology: The new synthesis*. Cambridge, MA: Harvard University Press.

Wolf, M., van Doorn, G. S., & Weissing, F. J. (2008). Evolutionary emergence of responsive and unresponsive personalities. *Proceedings of the National Academy of Sciences of the United States of America, 105*, 15825–15830.

CHAPTER 2

Assessment of sensory processing sensitivity across the lifespan

Francesca Lionetti
Department of Neurosciences, Imaging and Clinical Sciences, G. d'Annunzio University of Chieti-Pescara, Chieti, Italy
Department of Biological and Experimental Psychology, Queen Mary University of London, London, United Kingdom

Contents

1 SPS in the context of the environmental sensitivity metaframework 18
  1.1 SPS and phenotypic markers of sensitivity 19
2 Measures for the assessment of SPS in childhood and adolescence 22
  2.1 The Highly Sensitive Person—Parent-report scale 23
  2.2 The Highly Sensitive Child—Self-report scale 23
  2.3 The Highly Sensitive Child—Parent-report scale 27
  2.4 The Highly Sensitive Child Rating System 28
3 Measures for the assessment of SPS in adulthood 30
4 Dandelions, Tulips, and Orchids: A metaphor of individual differences in sensitivity 36
5 Future directions 41
References 44

A core aspect of individuals high in sensory processing sensitivity (SPS) is a heightened sensitivity to the influence of both positive and negative external environments, as well as an increased sensitivity to internal stimuli (Aron, Aron, & Jagiellowicz, 2012; Greven et al., 2019). This core mechanism of sensitivity is what the trait of SPS describes at a phenotypical level referring to greater depth of processing. For identifying individuals high in SPS, first a self-report scale for adults was developed (Aron & Aron, 1997). Subsequently, self-report, parent-report, and observational measures have been developed for children and adolescents. In the first section of this chapter, we introduce SPS as a measure of environmental sensitivity, reviewing empirical evidence showing that SPS is associated with a heightened susceptibility to stimuli. In Section 2, we introduce self-report, parent-report, and observational measures for the assessment of SPS in children and adolescents. In Section 3, we describe the Highly Sensitive Person (HSP) scale for the assessment of sensitivity in adults. Also, we include a discussion of the sensitivity scales'

https://doi.org/10.1016/B978-0-12-818251-2.00002-3

psychometric properties, construct validity, and context of applications across countries and developmental periods. In Section 4, we introduce the notion of sensitivity groups based on empirical findings suggesting the existence of three sensitivity categories. Finally, strengths and limitations of measures for the assessment of SPS, and new directions of study, are discussed.

1 SPS in the context of the environmental sensitivity metaframework

In agreement with the environmental sensitivity metaframework (Pluess, 2015) individuals differ in their sensitivity to the environment, with a minority of them, around 30% as recently estimated based on empirical data (Pluess et al., 2018; Lionetti et al., 2018) being *highly* sensitive. Traditionally, the most extensively studied phenotypical markers of heightened sensitivity have been negative emotionality and difficult temperament, and thus sensitivity has been considered synonymous with vulnerability. According to the Diathesis-stress model (Monroe & Simons, 1991), individuals scoring high on negative emotionality and/or difficult temperament show an increased susceptibility to negative environments with a dual-risk effect, such that the risk for negative outcomes is higher in the presence of risky conditions at both the individual (negative emotionality/difficult temperament) and the environmental (e.g., harsh parenting) level. Conversely, those scoring low on these traits are resilient in the face of adversities, but they do not benefit more from positive environments: they are overall not sensitive to environmental influences. The inclusion of negative emotionality, particularly in developmental studies, as a candidate phenotypic marker of an increased sensitivity to stimuli has been most likely the heritage of a developmental psychology approach that was largely interested in understanding the mechanisms underlying maladaptive outcomes, and thus predominantly focused on the adverse effects of problems and disorders (Rutter & Sroufe, 2000). As soon as the focus of studies started to include positive environments and positive outcomes, research evidence proved negative emotionality in infants to be a marker of sensitivity to positive environments as well. As such, children with a highly negative temperament showed evidence of outperforming their peers scoring low on this trait in terms of positive developmental outcomes when exposed to supportive environmental conditions, such as positive parenting (Belsky & Pluess, 2009; Kim & Kochanska, 2012). This increased sensitivity, for better and for worse, is what is postulated within the differential susceptibility model (Belsky & Pluess, 2009;

Belsky, Bakermans-Kranenburg, & Van IJzendoorn, 2007) and biological sensitivity to the context theory (Boyce & Ellis, 2005). The bright side of these models is captured by the more recent Vantage Sensitivity model (Pluess & Belsky, 2013), which postulates that some individuals are particularly susceptible to exceptionally positive environments, while not necessarily being more sensitive to negative stimuli. Though these theories differ from one another in predicting the type of response to the environment (and for other theoretical aspects described more in details in Pluess, 2015), these models agree on the notion that only a minority of the population presents an increased sensitivity to environmental and internal stimuli, while the majority is low in sensitivity. For this reason, they have been considered part of the broader metaframework of environmental sensitivity (Pluess, 2015).

Among theories of environmental sensitivity, there is also the term "sensory processing sensitivity" (Aron et al., 2012; Acevedo et al., 2014; Greven et al., 2019), which is the only theoretical perspective providing a range of direct measures for assessing sensitivity at a phenotypical level, and is described extensively in this edition. Research studies showed that SPS correlates with differential outcomes depending on the environment experienced: individuals high in SPS are at a greater risk for maladjustment in negative environments than individuals low in SPS (e.g., see Andresen, Goldmann, & Volodina, 2018), but they also profit more from positive environmental stimuli. For example, correlational studies showed that those with high SPS and poor childhoods were at an increased risk for depression, but in the absence of such negative conditions they are no more likely than others to be depressed (Aron, Aron, & Davies, 2005; Liss, Timmel, Baxley, & Killingsworth, 2005). Individuals high in SPS have also been reported to benefit more than others from positive stimuli, as shown in laboratory studies with positive mood-induction tasks and intervention studies (antibullying intervention and depression prevention program) where the environmental, independent variable was manipulated (Lionetti et al., 2018; Nocentini, Menesini, & Pluess, 2018; Pluess & Boniwell, 2015). This will be discussed further in Chapter 4 of this edition on outcomes of SPS.

1.1 SPS and phenotypic markers of sensitivity

At a phenotypical level, aspects of an increased sensitivity have long been observed in temperament and personality traits including inhibition, negative emotionality, and neuroticism. These traits, to some extent, have been shown to correlate with SPS across children and adults, though, importantly, they do not completely overlap with SPS (Pluess et al., 2018; Lionetti et al., 2018; Lionetti, Pastore, et al., 2019). Metaanalytical findings on the association between SPS and the Big Five personality traits (Lionetti, Pastore, et al., 2019) reported that in children SPS correlated with Neuroticism ($r = .42$) but did not with Extraversion, Openness, Agreeableness, or Conscientiousness. In adults, SPS correlated with Openness (though to a low extent, $r = .14$) and Neuroticism ($r = .40$) but did not with Extraversion, Agreeableness, or Conscientiousness.

Several theories and dimensions of temperament and personality have been proposed to capture individual differences, and what characterizes these different theories is that almost all distinguish between more and less reactive (responsive) individuals. In the developmental psychology theory (Kagan, 1989; Kagan, Reznick, & Snidman, 1987), when describing children's temperamental differences, distinctions have been made between inhibited/reactive (the more sensitive ones) and disinhibited/nonreactive, with the former presenting a strong theoretical overlap with the concept of SPS. Similarly, Gray's (1981, 1982) personality theory, in adults, distinguished between individuals high in the behavioral inhibition system (BIS) vs. those scoring high in the behavioral activation system (BAS). Though Gray initially defined the BIS system as linked to brain structures associated with anxiety and response to threatening stimuli, in the last 20 years the BIS has been more properly described as implying a higher inhibition before approaching or withdrawing due to a deeper processing of information (McNaughton & Gray, 2000). Again, this is something, which is captured also by SPS, whose core feature is a heightened depth of processing (Aron & Aron, 1997). This depth of processing is associated with an increased sensitivity to external and internal stimuli. As environmental events are perceived and processed more deeply by highly sensitive individuals, these individuals are also more affected for better and for worse, as shown by measures of outcomes and adjustment as discussed in Chapter 4 (see, for example, the perception of stress at work in highly sensitive adults in Evers, Rasche, & Schabracq, 2008, and the positive response to intervention programs in Pluess & Boniwell, 2015 and Nocentini et al., 2018).

Together with inhibition, negative emotionality has been another widely investigated marker of increased sensitivity to the environment, particularly in children (for a metaanalysis, see Slagt, Dubas, Dekovic, & van Aken, 2016). Because negative emotionality involves the expression of negative emotions in response to negative environmental conditions (Rothbart & Bates, 2006), it may appear intuitive that displays of negative emotionality may serve as a marker of high sensitivity, particularly in adverse environments. Also, highly sensitive children may feel easily overwhelmed when exposed to negative stimuli (such as chaotic and loud environments, bright lights, and strong noises), thus expressing behavioral reactions such as crying and protesting. In adults, for example, high SPS has been shown to be associated with increased sensitivity to chaotic and loud home environments. Empirical evidence showed that associations between higher home-crowding and perceptions of home chaos were significant only for highly sensitive mothers (Wachs, 2013). However, high negative emotionality has been reported as a marker of increased sensitivity to positive stimuli too (for a metaanalysis, see Slagt et al., 2016). As such, researchers have described the trait more generally as "reactivity," "responsiveness," and high sensitivity with corresponding theories about differential susceptibility (Belsky & Pluess, 2009; Belsky et al., 2007), biological sensitivity to context (Ellis & Boyce, 2008) and sensory processing sensitivity (Aron & Aron, 1997; Aron et al., 2012), which will be discussed in detail in Chapter 3. According to these theories of environmental sensitivity, negative emotional reactions reflect an increased sensitivity of the central nervous system to even mildly negative stimuli, which more broadly is associated with both positive and negative events being registered more easily (Pluess & Belsky, 2010).

However, recent empirical evidence pointed out that negative emotionality is a strong marker of an increased sensitivity to the environment in infancy and toddlerhood, but does not seem to capture sensitivity toward both positive and negative environments at an older age (Slagt et al., 2016). From around preschool years, children scoring high on this trait have been reported to be rather more susceptible to the influence of negative environments only. One candidate explanation is that if raised by a sensitive and responsive parent, a child who in early infancy was high in negative emotionality (and highly sensitive) may express a decrease in his/her negative emotional reactions. Partly, this may be due to lower frequency of exposure to negative stimuli, due to greater supporting, nurturing, and positive environmental experiences. Alternatively, and perhaps correspondingly, in such types of supportive and positive environments, the child may have learned to better regulate emotions, as a result of positive parenting practices (Karreman, Van Tuijl, van Aken, & Deković, 2006). In fact, there is evidence with adult brain imaging studies showing that when exposed to positive or negative pictures, highly sensitive individuals reporting positive childhood environments, confer stronger brain activity in regions that are involved in self-regulation (Acevedo, Jagiellowicz, Aron, Marhenke, & Aron, 2017). This will be discussed in greater detail in forthcoming chapters.

One limitation in considering negative emotionality and behavioral inhibition as selective markers of an increased sensitivity is that both are, to some extent, biased by a focus on negative aspects only, and may fail to phenotypically reflect sensitivity in response to positive and pleasant stimuli too. Though SPS does indeed correlate with neuroticism and negative affect (Aron et al., 2012; Lionetti, Pastore, et al., 2019), which capture the negative features of SPS, individuals scoring high on SPS likewise show stronger positive emotional reactions when exposed to positive stimuli (Lionetti et al., 2018; Nocentini et al., 2018). Also, compared to negative emotionality and inhibition, SPS also captures greater depth of processing, which is not limited to negative stimuli. For example, highly sensitive individuals have been proposed to be more sensitive to the arts (Bridges & Schendan, 2019). How the construct of SPS has been operationalized through a series of phenotypic measures for directly capturing sensitivity will be summarized in the forthcoming paragraphs.

2 Measures for the assessment of SPS in childhood and adolescence

In the 1990s, Aron and Aron conducted an empirical study (Aron & Aron, 1997) that led to the development of the 27-item Highly Sensitive Person scale, adopted since then for the assessment of SPS in adults. Items composing this scale were identified through a series of interviews with persons self-identified as "highly sensitive" in order to gain a better understanding of what sensitivity is. The scale resulted in measurement of the cardinal features of SPS such as being easily overwhelmed by stimuli (e.g., *I am easily overwhelmed by things like bright lights, strong smells, coarse fabrics, or sirens close by*; *I get rattled when I have a lot to do in a short amount of time*), and showing a higher appreciation of, and a stronger emotional response to, positive stimuli (e.g., *I am deeply moved by the arts or music*). As such, this variety of items (correlating with each other) was capturing a broader phenomenon than being sensitive to sensory stimuli only, or being shy or inhibited. However, some core features of the theoretical definition of SPS, such as positive emotional reactivity and depth of processing, are not fully covered by items originally included in the HSP scale. As such, the development of additional items able to capture depth of processing and positive emotionality features of SPS is currently underway by the authors of the HSP Scale.

Since the publication of the paper presenting the HSP scale in 1997, SPS has been studied in personality research studies with adults (for a metaanalysis see Lionetti, Pastore, et al., 2019). This has been probably due also to the rapid dissemination of the HSP scale as a measure for capturing SPS in adulthood.

2.1 The Highly Sensitive Person—Parent-report scale

In more recent years, the study of SPS has started to capture the attention of developmental psychology too, and the number of studies investigating SPS in children has significantly increased thanks to the development and validation of scales for phenotypically assessing SPS. Inspired by the HSP Scale for adults, the first measure designed for directly capturing SPS in children was the 23-item parent-report HSP questionnaire (Aron, 2002). Proposed in 2002 in the "Highly Sensitive Child" book by Elaine Aron in a "yes" vs "no" answer format, it has been only recently empirically tested for its association with a set of candidate symptoms, and for its factorial structure, in a study featuring a group of 253 parents based in Flanders, Belgium. Parents were asked to report on their 3- to 16-year-old child's sensitivity using the 23-item parent-report HSP questionnaire, rating each item on a 5-point Likert scale. Parents were also asked to report on children's medically unexplained symptoms in daily functioning with a set of questions developed ad hoc, investigating sleeping problems, abdominal and headache pain, and drinking and eating problems. Additionally, the study explored the presence of (and SPS' association with) antisocial behaviors (e.g., stealing, bullying). Results from this study showed that the parent-report HSP scale captured two factors: Overreaction to Stimuli, which comprised items associated with overstimulation, emotional intensity, and sensory sensitivity (e.g., "my child is bothered by loud noises"); and Depth of Processing (e.g., "my child asks deep, thought-provoking questions"). Association with symptom variables showed that children with more problems in daily functioning (i.e., sleeping, eating, and drinking problems, crying excessively) scored higher on SPS, specifically on the Overreaction to Stimuli factor, especially for what pertains to crying excessively, whereas Depth of Processing was mainly related with sleeping problems (Boterberg & Warreyn, 2016). However, the extent to which the HSP parent-report scale captures an increased sensitivity to both positive and negative stimuli has yet to be investigated.

2.2 The Highly Sensitive Child—Self-report scale

Another measure for the assessment of SPS in children and adolescents is the 12-item Highly Sensitive Child (HSC) self-report scale (Pluess et al., 2018, Table 1). This scale is also available in a parent-report format for investigating

Table 1 The Highly Sensitive Child scale.

Items	Factors[a]
1. I notice when small things have changed in my environment	Aesthetic Sensitivity
2. Loud noises make me feel uncomfortable	Low Sensory Thresholds
3. I love nice smells	Aesthetic Sensitivity
4. I get nervous when I have to do a lot in little time	Ease of Excitation
5. Some music can make me really happy	Aesthetic Sensitivity
6. I am annoyed when people try to get me to do too many things at once	Ease of Excitation
7. I don't like watching TV programs that have a lot of violence in them	Low Sensory Thresholds
8. I find it unpleasant to have a lot going on at once	Ease of Excitation
9. I don't like when things change in my life	Ease of Excitation
10. I love nice tastes	Aesthetic Sensitivity
11. Loud noises make me feel uncomfortable	Low Sensory Thresholds
12. When someone observes me, I get nervous. This makes me perform worse than normal	Ease of Excitation

[a] Factors as reported in Pluess et al. (2018).

sensitivity in preschoolers (Slagt, Dubas, van Aken, Ellis, & Dekovic, 2018). Initially proposed in a 7-point Likert-type format, the HSC self-report scale has also been used in a 5-point format for younger children (e.g., see Nocentini et al., 2018), with pupils from grades 4 to 6. The self-report HSC scale has been extensively explored for its psychometric proprieties in a UK-based validation study involving three independent children samples (for a total of 696 participants), and one sample of 1431 adolescents (Pluess et al., 2018) ages 9–16 years. Results showed that the scale is psychometrically robust, and has a bifactor structure as reflected in a general sensitivity factor, and also in three specific factors capturing the perception of being overwhelmed by both external and internal stimuli (Ease of Excitation factor—EOE, e.g., being negatively affected by having a lot going on around, or getting nervous when observed); aesthetic awareness and sensitivity to positive stimuli (Aesthetic Sensitivity factor—AES, e.g., appreciation of nice tastes and nice smells); and Low Sensory Thresholds in response to potentially disturbing sensory experiences (Low Sensory Thresholds factor—LST, e.g., being disturbed by loud noises and violent TV programs). Interestingly, correlation and regression analyses across samples showed that the scale is relatively independent from other common personality and temperament traits, which explained up to 34% of variance for the HSC summary score, depending on the sample and on the specific measures adopted for the concurrent assessment of temperament and personality (Pluess et al., 2018). The HSC scale thus captures a specific sensitivity factor, not otherwise captured by other existing personality and temperament measures, as also reported in a recent metaanalysis (Lionetti, Pastore, et al., 2019).

This brief self-report HSC child scale inspired by the literature on SPS does capture sensitivity to the environment in children, as it has been reported across two intervention studies and in one correlational study. Results from Pluess and Boniwell (Pluess & Boniwell, 2015) with a sample of 166, 11-year old girls from an at-risk population in England, undergoing a school-based prevention depression program, showed that pupils scoring high on SPS were most positively affected by the intervention, compared to less sensitive students. More specifically, depression scores, assessed using the widely known Center for Epidemiologic Studies Depression Scale (Radloff, 1977), continued to decrease significantly among highly sensitive females, even 6 and 12 months after completing the intervention. These results are consistent with SPS as a marker of an increased environmental sensitivity and, more specifically, are in line with the Vantage Sensitivity model. That is, highly sensitive individuals were most susceptible to the influence of positive stimuli, outperforming their low-sensitive peers (Pluess & Belsky, 2013). On the contrary, depression rates remained stable in those scoring low on the HSC scale. We can speculate that the more sensitive brain of pupils scoring high on the 12-item HSC scale allowed the group undergoing the prevention program to process information related to the intervention more deeply, thus leading to a greater reduction of depression rates. While we still need to clarify the mechanism(s) through which greater SPS would lead to a greater reduction of symptoms in response to an intervention program, something which was not been investigated in the mentioned study, we can nonetheless conclude that the HSC self-report scale is able to capture an increased sensitivity to a positive stimulus in intervention program in a group of adolescents. Thus, in addition to showing the positive side of SPS, this study also showed evidence that the HSC scale is a valid measure for the assessment of sensitivity to environmental stimuli.

This sensitivity response to exceptionally positive stimuli has been recently reported also in a large scale randomized-controlled trial aimed at testing a school-based antibullying intervention in an Italian-based population (Nocentini et al., 2018). The study involved 2042 nine- to ten-year-old childrenrandomly divided into two groups: the intervention group and the control (no-intervention) group. Children completed a battery of questionnaires, including the HSC self-report scale, a scale used to measure how often respondents had experienced bullying behaviors as victims or perpetrators (Palladino, Nocentini, & Menesini, 2015), and the Youth Self-Report questionnaire for investigating internalizing and externalizing symptoms (Achenbach, 1991). Results showed that significant intervention effects on victimization and internalizing symptoms were driven by gender and sensitivity. That is, boys scoring high on the HSC self-report scale benefitted significantly more from intervention than their less sensitive peers. Interestingly, boys scoring high on the HSC scale and assigned to the intervention condition, showed a reduction of victimization with an effect size that was double to that of the medium sensitive group. On the other hand, children scoring low on the HSC scale showed no improvement at all in response to the intervention program. The study reported comparable results for the internalizing symptoms domain.

Moderation effects of SPS on children's adjustment were also reported in a recent correlational study involving 277 children (age range: 5–12 years, mean age: 7.05, SD = 1.01), where the HSC scale was administered in an interview format, with the experimenter reading each item loudly to the child (Scrimin, Osler, Pozzoli, & Moscardino, 2018). Children and their parents were interviewed also on experienced childhood adversities (as reported by parents using an ad hoc checklist of stressful events), on supportive resources in the family environment, and on children's well-being using the Child Health and Illness Profile (Riley et al., 2004, 2004). Results showed that children scoring high on the HSC scale were more susceptible, than their less sensitive peers, to the negative influence of a stressful environment and also to the positive influence of perceived family support (Scrimin et al., 2018). Specifically, for highly sensitive children, a greater number of childhood family adversities was associated with lower levels of physical comfort and less social competence. Similarly, more supportive resources provided by the family were associated with better social competence in children scoring high on the HSC scale. On the contrary, when sensitivity was low, the effect of the environment on outcomes was significantly lower or even not significant, depending on the variable investigated.

Overall, these studies suggest, across independent samples, that the self-report HSC scale is able to capture an increased sensitivity to both positive and negative stimuli, and thus it may be considered a reliable measure for the assessment of SPS in school-aged children and adolescents.

2.3 The Highly Sensitive Child—Parent-report scale

Recently, the HSC self-report scale (Pluess et al., 2018) has been adapted in a parent-report format (Slagt et al., 2018), including the same items used in the HSC self-report scale but with the parent as the main informant of the child's sensitivity (e.g., "Loud noises make me uncomfortable" in the HSC self-report scale is rephrased as "Loud noises make my child uncomfortable" in the HSC parent-report scale, Table 2). Compared to the 23-item parent-report HSP scale, the HSC-parent report includes items that tap into three different factors, the same reported for the HSC/HSP self-report scales (i.e., Ease of Excitation, Aesthetic Sensitivity, and Low Sensory Thresholds).

A study using the HSC parent-report measure provided support for this scale's effectiveness in capturing a child's sensitivity to the environment (Slagt et al., 2018). The study involved 264 Dutch kindergartners followed for three waves, spaced seven months apart. Teachers reported on each child's externalizing behaviors and prosocial behaviors using the Strengths and Difficulties Questionnaire (Goodman, 1997). Parents reported on their children's sensitivity and on their own parenting using a series of scales for investigating positive and negative parenting (Slagt et al., 2018). The analysis of longitudinally collected data showed that SPS, as reported by parents with the HSC parent-report scale, interacted with both (changes in) negative and (changes in) positive parenting in predicting externalizing behavioral problems. Interestingly, negative emotionality, another widely explored marker of sensitivity already discussed in this chapter, did not moderate the impact of the environment on outcomes, thus suggesting that, at least from preschool years, SPS may be a more proximal and accurate marker of individual differences in response to the environment compared to the widely investigated negative emotionality trait. Results support that through a relatively brief, and psychometrically valid questionnaire, such as the HSC parent-report scale, it is possible to adequately measure an increased sensitivity to environmental influences, and more specifically interpersonal relationships, in children.

Starting from the original English version, the HSC self-report scale has been translated into Dutch (Weyn et al., 2019), Italian (Nocentini, Menesini, Lionetti, & Pluess, 2017), Spanish (Silvestre, 2018), Hebrew (Goldberg et al., 2018). Partial measurement invariance has been supported across United Kingdom and Dutch samples (Weyn et al., 2019).

Table 2 The Highly Sensitive Child scale—parent report.

Items	Factors[a]
1. My child notices when small things have changed in his/her environment	Aesthetic Sensitivity
2. Loud noises make my child feel uncomfortable	Low Sensory Thresholds
3. My child loves nice smells	Aesthetic Sensitivity
4. My child gets nervous when he/she has to do a lot in little time	Ease of Excitation
5. Some music can make my child really happy	Aesthetic Sensitivity
6. My child gets annoyed when people try to get him/her to do too many things at once	Ease of Excitation
7. My child doesn't like watching TV programs that have a lot of violence in them	Low Sensory Thresholds
8. My child finds it unpleasant to have a lot going on at once	Ease of Excitation
9. My child doesn't like when things change in his/her life	Ease of Excitation
10. My child loves nice tastes	Aesthetic Sensitivity
11. My child doesn't like loud noises	Low Sensory Thresholds
12. When someone observes my child, he/she get nervous. This makes my child perform worse than normal	Ease of Excitation

[a] Factors as reported in Slagt et al. (2018).

2.4 The Highly Sensitive Child Rating System

The questionnaires introduced thus far have several strengths. One is that they require very limited administration time, allowing them to be completed along with other measures. Also, the parent-report HSP, self-report HSC, parent-report HSC all have strong psychometric proprieties. The scales can be completed by parents of preschoolers (parent-report HSP, parent-report HSC), and do capture an increased sensitivity to stimuli, moderating the impact of the environment on a series of assessed outcomes (self-report HSC and parent-report HSC). However, parents might not always be reliable observers of their children's behavior (as, for example, in samples of at-risk families), or might not be available to complete questionnaires. For overcoming these limitations, recently an observational measure for the assessment of SPS in preschool children has been proposed: the HSC rating system. This rating system represents the first attempt to directly capture sensitivity at an observational level and has its roots in SPS and in the environmental sensitivity theories (Pluess, 2015). The HSC rating system consists of 10 scales (Lionetti, Aron, Aron, Klein, & Pluess, 2019, Table 3)

Table 3 The Highly Sensitive Child Rating System (HSC-RS).

LabTAB episodes	HSC-RS Scales[a]
Risk room	1. Pause to check before exploring a new environment 2. Cautious and collaborative attitude toward the experimenter
Tower of patience	3. Attending to experimenter's directions
Stranger approach	4. Compliance with the experimenter's request 5. Fearfulness in response to the stranger's entrance
Exploring new objects	6. Hesitancy paired with curiosity
Pop-up snakes	7. Positive response/overexcitement
Transparent box	8. Attention to toys' detailed features 9. Careful perseverance when trying to open the box
Impossibly perfect children	10. Preference for (and commitment to drawing) beautiful circles

[a] As reported in Lionetti, Aron, et al. (2019).

applied to a series of laboratory episodes derived from the Laboratory Temperament Assessment Battery procedure (Lab-TAB), traditionally used for coding temperament (Goldsmith et al., Unpublished manuscript). Data based on a sample of 292 children in the United States (Lionetti, Aron, et al., 2019) showed that the measure fits a one-factor solution, and that this unique sensitivity factor does not completely overlap with the combined contribution of observed temperament factors (i.e., Sociability, Positive Affect/Interest, Dysphoria, Fear/Inhibition, and Constraint vs Impulsivity) as assessed with the Lab-TAB (Dyson, Olino, Dunn, Goldsmith, & Klein, 2012). The measure has been shown to capture sensitivity to both positive and negative quality parenting. Children scoring high on the sensitivity summary score, being more vulnerable to the effects of negative parenting (and, specifically, high levels of permissive parenting) experienced at age 3, showed increased externalizing behavioral problems at age 3 and internalizing behavioral problems at ages 3 and 6. Also, the same children showed the highest levels of social competence in response to positive parenting (high levels of authoritative parenting), and results remained stable after controlling for negative affect (Lionetti, Aron, et al., 2019). Overall, these research findings suggest that this newly developed observational measure is able to capture an increased sensitivity to stimuli and provides an opportunity for exploring sensitivity in young children. Combined with parent-report scales, it allows a multiinformant and multiassessment investigation of sensitivity in young children.

Importantly, the availability of measures inspired by the same theoretical background, but covering different age ranges, potentially allows for researchers to study sensitivity longitudinally, something which has not been investigated thus far. Though sensitivity is theoretically considered to be a trait that is largely genetically determined, according to some developmental theories (Ellis & Boyce, 2008) the environment might play a role in shaping individual differences in sensitivity. Assessing the degree of change and stability of sensitivity across human (and perhaps even animal) development would allow us to more extensively test this hypothesis.

Finally, for assessing sensitivity among children in the school context, currently a teacher-report and an observational measure for primary school children are under development and validation in an ongoing 2-year longitudinal study (Pluess & Lionetti, 2018). The aim of this study is to extend examination of sensitivity to the educational field, and more specifically to understand how individual differences interact with the quality of the learning environment in promoting healthy development and academic achievement. The teacher-report scale currently includes a series of items rated on a 7-point Likert-scale evaluating children's sensitivity to sensory stimuli (e.g., struggling to focus in chaotic environments) and interpersonal relationships (e.g., being deeply affected by criticisms and failure; sensitivity to praises; sensitivity to injustice), as shown in the school context.

Published studies investigating SPS in children and adolescents using measures specifically developed for the assessment of SPS, as reviewed in the current paragraph, are summarized in Table 4.

3 Measures for the assessment of SPS in adulthood

SPS theory initially attracted the attention of social, clinical, and personality researchers and clinicians working with adults. Because the SPS trait was first empirically explored in adults, this result might be expected. It was 1997 when Aron and Aron published the first exploratory and empirical study aimed at uncovering what they initially defined as a general increased sensitivity and responsivity in humans. The authors recruited, through advertisements, 39 adults self-identified as highly sensitive, introverted, or easily overwhelmed by stimuli, and performed a series of interviews for identifying key markers of increased sensitivity to external and internal stimuli (Aron & Aron, 1997). Through this interview process, Aron and Aron identified a series of behavioral patterns leading to the identification of 60 items that converged into a first draft of the Highly Sensitive Person

Table 4 Empirical studies involving children and adolescents and in which SPS has been assessed using the Highly Sensitive Child or the Highly Sensitive Person self-report/parent-report measures, or the HSC Rating System.

Author(s)	SPS measure	Country	Age (in years)	Variables—other than SPS—investigated
Pluess and Boniwell (2015)	HSC self-report	United Kingdom	11	Depression
Boterberg and Warreyn (2016)	HSP-parent report	Belgium	3–16	Medical unexplained physical symptoms
Nocentini et al. (2017)	HSC self-report	Italy	9–12	Temperament
Slagt et al. (2018)	HSC parent-report	Netherlands	3–7	Positive and negative parenting, child's externalizing and prosocial behavior
Goldberg et al. (2018)	HSP self-report	Israel	16	Type I Diabetes
Pluess et al. (2018)	HSC self-report	United Kingdom	8–19	Temperament and personality traits
Nocentini et al. (2018)	HSC self-report	Italy	9–12	Bullying, victimization, internalizing and externalizing symptoms.
Scrimin et al. (2018)	HSC self-report	Italy	5–12	Family adversities, children's physical and emotional comfort, perceived academic performance, child's well-being
Tillmann, El Matany, and Duttweiler (2018)	HSP self-report for children	Germany	16	Personality, academic interest, adjustment and well-being, school performance
Weyn et al. (2019)	HSC self-report	Belgium	12–16	Temperament and personality traits
Lionetti, Aron, et al. (2019)	HSC rating system	United States	3	Parenting styles, externalizing and internalizing behavioral problems, social competence

scale. The 60-item questionnaire included statements capturing a broader sensitivity concept as being highly sensitive to others' emotions and needs, high sensitivity to external and internal stimuli, presenting lower sensory thresholds and having a rich inner life. This first 60-item version was tested on a relatively large sample of over 900 undergraduate psychology students and individuals from a community sample. Results from the factor analysis resulted in the 27-item HSP scale (Table 5, Aron, 1996), currently used for assessing SPS in adults (Aron & Aron, 1997).

Table 5 The 27-item Highly Sensitive Person scale.

Items	Factors[a]
1. Are you easily overwhelmed by strong sensory input?	Low Sensory Thresholds
2. Do you seem to be aware of subtleties in your environment?	Aesthetic Sensitivity
3. Do other people's moods affect you?	Ease of Excitation
4. Do you tend to be more sensitive to pain?	Ease of Excitation
5. Do you find yourself needing to withdraw during busy days, into bed or into a darkened room or any place where you can have some privacy and relief from stimulation?	Aesthetic Sensitivity
6. Are you particularly sensitive to the effects of caffeine?	Low Sensory Thresholds
7. Are you easily overwhelmed by things like bright lights, strong smells, coarse fabrics, or sirens close by?	Low Sensory Thresholds
8. Do you have a rich, complex inner life?	Aesthetic Sensitivity
9. Are you made uncomfortable by loud noises?	Low Sensory Thresholds
10. Are you deeply moved by the arts or music?	Aesthetic Sensitivity
11. Does your nervous system sometimes feels so frazzled that you have to go off by yourself?	Ease of Excitation
12. Are you conscientious?	Aesthetic Sensitivity
13. Do you startle easily?	Ease of Excitation
14. Do you get rattled when you have a lot to do in a short amount of time?	Ease of Excitation
15. When people are uncomfortable in a physical environment do you tend to know what needs to be done to make it more comfortable (like changing the lighting or the seating)?	Aesthetic Sensitivity
16. Are you annoyed when people try to get you to do too many things at once?	Ease of Excitation
17. Do you try hard to avoid making mistakes or forgetting things?	Ease of Excitation

Table 5 The 27-item Highly Sensitive Person scale—cont'd

Items	Factors[a]
18. Do you make a point to avoid violent movies and TV shows?	Low Sensory Thresholds
19. Do you become unpleasantly aroused when a lot is going on around you?	Low Sensory Thresholds
20. Does being very hungry create a strong reaction in you, disrupting your concentration or mood?	Ease of Excitation
21. Do changes in your life shake you up?	Ease of Excitation
22. Do you notice and enjoy delicate or fine scents, tastes, sounds, works of art?	Aesthetic Sensitivity
23. Do you find it unpleasant to have a lot going on at once?	Ease of Excitation
24. Do you make it a high priority to arrange your life to avoid upsetting or overwhelming situations?	Ease of Excitation
25. Are you bothered by intense stimuli, like loud noises or chaotic scenes?	Low Sensory Thresholds
26. When you must compete or be observed while performing a task, do you become so nervous or shaky that you do much worse than you would otherwise?	Ease of Excitation
27. When you were a child, did your parents or teachers seemed to see you as sensitive or shy?	Ease of Excitation

Note: In bolds are items from the 12-item HSP scale (Pluess, Lionetti, Aron, & Aron, 2020).

[a] Factors as reported in Smolewska, McCabe, and Woody (2006), except for items 1 and 11, removed by Smolewska et al. (2006), and included in Lionetti et al. (2018). For a discussion on the existence of a general sensitivity factor in adults, see Lionetti et al. (2018).

Since the development of the HSP scale, a widely debated aspect has pertained to its factor structure (Aron & Aron, 1997). The first psychometric analysis of the factor structure of the scale suggested a one-factor solution in spite of the variability of the items' content. Later on, the scale was used across several behavioral and brain activity studies, which provided empirical evidence that the score resulting from the sum of items of the Highly Sensitive Person scale correlates with an increased depth of processing and increased emotional reactivity to both positive and negative environmental contexts (Acevedo et al., 2014; Aron et al., 2005, 2012; Jagiellowicz et al., 2012; Lionetti et al., 2018).

However, subsequent exploratory factor analyses investigating alternative candidate solutions found convergence for different components (Smolewska et al., 2006). In Western samples, up until recently, one of the most extensively reported factor solutions includes the following: (1) Ease of Excitation (EOE, namely, being easily overwhelmed by internal and external stimuli), (2) Aesthetic sensitivity (AES, i.e., openness for, and pleasure of, aesthetic experiences and positive stimuli), (3) Low Sensory Threshold (LST, capturing sensitivity to external stimuli as bright lights or loud noises), as previously introduced in relation to the HSC child-report and parent-report scales (Pluess et al., 2018). Across several independent studies, these three sensitivity components of EOE, AES, and LST have been found to relate differentially to negative and positive affect variables. More specifically, both Low Sensory Threshold (LST) and Ease of Excitation (EOE) were found to be associated to a moderate extent with self-reported negative emotionality, anxiety, and depression, but not EOE (for a review of studies investigating HSP factors and affect variables see Greven et al., 2019). Conversely, the component reflecting the more positive, bright responses of sensitivity toward positive stimuli, namely, Aesthetic sensitivity (AES), was found to correlate with positive affect and self-esteem, but not with negative emotions (Liss et al., 2005; Smolewska et al., 2006). However, it is not clear what these three components do actually capture when considered separately, and the relevance of these three factors when considered separately has yet to be determined. Their identification resulted from a post hoc analysis of the HSP adult questionnaire, and these components have been subsequently adopted as a guide for orienting the identification of the HSC scale items. Theoretically, another important thing that has to be considered when looking at the three sensitivity factors is that currently none of these components explicitly captures depth of processing, a key feature of SPS. To some extent, some of the items currently included in the Aesthetic Sensitivity factor do capture an increased depth of processing (e.g., see *do you seem to be aware of subtitles in the environment?*), but these items are indirect and few.

Currently, a team of researchers is in the process of identifying and validating an additional set of items to better capture depth of processing with the HSP scale. Another aspect that requires attention when considering the EOE, AES, and LST factor solutions is that the factor structure of the HSP scale has been theoretically proposed to capture a unitary sensitivity factor, and the relevance of these three factors when considered separately has yet to be determined. Indeed, empirical studies have most often considered a composite score as derived from the HSP scale when studying associations with positive and negative outcomes.

Recently, a way for reconciling the competitive three-factor model with the originally hypothesized one-factor structure has been proposed with a bifactor model. A bifactor model allows exploring the degree to which a measure yields a univocal total score as well as distinct subscales (Reise, Bonifay, & Haviland, 2013; Reise, Moore, & Haviland, 2010). Compared to the EOE, AES, and LST three-factor solution, in a bifactor model, specific factors are constrained to be orthogonal (i.e., uncorrelated), and each item is allowed to load both on a specific factor and on a general factor. In this way, the residual variance is not explained by the correlation among factors but is attributed to a general factor. Data currently available across childhood, adolescence, and adulthood provide support for a bifactor solution (Lionetti et al., 2018; Pluess et al., 2018). This solution includes a general SPS factor, and also recognizes the multidimensionality of the HSP scale as represented by the three EOE, AES, and LST sensitivity components (Lionetti et al., 2018; Pluess et al., 2018).

With comparable psychometric properties, a short 12-item version of the 27-item HSP scale has been proposed and utilized in research studies (see Table 5, items in bold). Data from a series of independent UK-based adults supported the 12-item HSP scale to have strong psychometric proprieties, to not completely overlap with the five factors of personality, and to capture an increased sensitivity toward both positive and stressful stimuli, as shown across multiple research conditions (correlational, longitudinal, laboratory-based; Pluess et al., 2020).

Currently, the HSP scale has been adopted with German (Konrad & Herzberg, 2017; Tillmann et al., 2018), Turkish (Şengül-İnal & Sümer, 2017), Japanese (Kibe, Suzuki, & Hirano, 2018), Icelandic (Þórarinsdóttir, 2018), Italian (Rubaltelli, Scrimin, Moscardino, Priolo, & Buodo, 2018), Iranian (Ghorbani Taghlidabad & Tasbihsazan Mashhadi, 2018), and South African samples (May, 2019). Invariance across countries has yet to be explored.

Research using the HSP scale with adult samples has linked SPS to a variety of maladaptive outcomes and psychopathology symptoms, partially due to a strong interest among researchers for uncovering the basis of maladjustment among high SPS individuals. These studies are discussed at length in Chapter 4 detailing the health and social outcomes of SPS. By understanding the factors underlying negative outcomes for SPS, researchers and program developers could then target prevention and intervention programs more efficiently. For example, SPS has been shown to be correlated with anxiety, depression, and poor social skills among college students

(Liss, Mailloux, & Erchull, 2008; Liss et al., 2005). Also, SPS in adults has been associated with lower levels of subjective happiness (Sobocko & Zelenski, 2015), life satisfaction (Booth, Standage, & Fox, 2015), increased levels of stress (Bakker & Moulding, 2012; Benham, 2006), poor stress management (Brindle, Moulding, Bakker, & Nedeljkovic, 2015), greater perception of home chaos (Wachs, 2013), physical symptoms of self-reported ill health (Benham, 2006), higher rates of turnover intentions in expatriate workers due to greater levels of stress (Andresen et al., 2018), greater levels of stress in the work environment (Evers et al., 2008), and irritable bowel syndrome (Ghorbani Taghlidabad & Tasbihsazan Mashhadi, 2018).

When research studies enlarged the focus toward the inclusion of positive environmental stimuli and positive outcomes, empirical evidence started to provide support for the notion that SPS also captures an increased sensitivity to positive stimuli. As reviewed, empirical studies have provided support for an increased sensitivity to positive interventions in children and adolescents scoring high on SPS, measured with the HSC scales (Nocentini et al., 2018; Pluess & Boniwell, 2015). Recently, more studies have started to investigate SPS in relation to positive variables in adult samples. For example, SPS (assessed with the HSP scale) has been shown to correlate with positive emotions after a positive mood manipulation laboratory task (Lionetti et al., 2018), creativity (Bridges & Schendan, 2019), and feelings of awe (Aron, Aron, & Tillmann, 2018; Shiota, Keltner, & Mossman, 2007).

Studies using the HSP scale with adult samples are summarized in Table 6.

4 Dandelions, Tulips, and Orchids: A metaphor of individual differences in sensitivity

That individuals differ with respect to their sensitivity level is one of the core assumptions of SPS, also for temperament and personality traits in general. From its origins, SPS theory has proposed SPS to be a dichotomous trait, with around 20% of the population found to be highly sensitive, and 80% being less sensitive (Aron et al., 2012). This proportion is mainly based on theoretical reasons and research evidence from other temperament traits which, as SPS, have been reported in association with an increased sensitivity to environmental influences (Woodward, Lenzenweger, Kagan, Snidman, & Arcus, 2000). The assumption that only a minority of the population is high on sensitivity also has its foundation in biological and evolutionary explanations, which suggest that a general sensitivity is an advantage for the

Table 6 Empirical studies on adults' SPS assessed with the HSP self-report scale.

Author(s)	Country of origin	Sample	Variables—other than SPS—investigated
Aron and Aron (1997)	United States	Undergraduates and community sample of convenience	Neuroticism
Meyer and Carver (2000)	United States	Undergraduates	Childhood memories, life orientation, mood, avoidant personality disorder
Neal, Edelmann, and Glachan (2002)	United Kingdom	Sample of convenience and participants of self-help and anxiety organizations	Phobia and anxiety, depression, inhibition,
Aron et al. (2005)	United States	Undergraduates	Adverse childhood environment and negative parenting, shyness, negative affectivity
Meyer, Ajchenbrenner, and Bowles (2005)		Sample of convenience of adults with borderline and avoidance features	Mood and pessimistic cognitive-affective reactions, personality disorder, attachment, childhood memory
Benham (2006)	United States	Undergraduates	Perceived stress and physical symptoms
Smolewska et al. (2006)	Canada	Undergraduates	Personality, behavioral inhibition, and behavioral activation system.
Hofmann and Bitran (2007)	United States	Outpatients, anxiety disorders	Personality, anxiety, social phobia
Evans and Rothbart (2008)	United States	Undergraduates	Adult temperament
Evers et al. (2008)	The Netherlands	Convenience sample of workers	Sense of coherence, self-efficacy, alienation, negative affectivity, work stress
Liss et al. (2008)	United States	Undergraduates	Anxiety, depression, alexithymia, autism
Kjellgren, Lindahl, and Norlander (2009)	Sweden	Undergraduates	Experience of relaxation in a flotation stimulation technique, pain endurance, anxiety, life orientation, stress
Aron et al. (2010)	United States	Convenience sample of East Asian and Americans	Neural responses to visual stimuli, personality, cultural specific identity

Continued

Table 6 Empirical studies on adults' SPS assessed with the HSP self-report scale—cont'd

Author(s)	Country of origin	Sample	Variables—other than SPS—investigated
Ahadi and Basharpoor (2010)	Iran	Undergraduates	Personality, mental and physical health
Woods, Colvin, Vera-Diaz, and Peli (2010)	United States	Undergraduates	Blur tolerance, personality
Chen et al. (2011)	China	Undergraduates	Perceived parental warmth, stressful life events, dopamine system
Bakker and Moulding (2012)	Australia	Community sample and undergraduates	Mindfulness, depression, anxiety and stress
Booth et al. (2015)	United Kingdom	Sample recruited through website	Childhood experiences and life satisfaction
Gearhart and Bodie (2012)	United States	Undergraduates	Stress; communication apprehension
Gerstenberg (2012)	Germany	Undergraduates	Personality, stress, visual detection task
Wachs (2013)	United States	Mothers	Environmental chaos, noise sensitivity, extraversion
Acevedo et al. (2014)	United States	Convenience sample of newly weds and engaged couples	Affective responses, neural activation
Gearhart (2014)	United States	Undergraduates	Neuroticism, distraction, nonverbal decoding accuracy
Jonsson, Grim, and Kjellgren (2014)			Nonordinary states of consciousness, depression, anxiety, optimism, absorption, experience of mystical state
Brindle et al. (2015)	Australia	Sample of convenience	Depression, anxiety, stress, emotion regulation, distress tolerance
Chen et al. (2015)	China	Undergraduates	Dopamine system, resting-state brain activity
Jagiellowicz, Aron, and Aron (2016)	United States	Undergraduates	Neuroticism, social extraversion, parenting quality, affect, arousal

Listou Grimen and Diseth (2016)	Norway	Undergraduates	Personality, subjective health complaints
Meredith, Bailey, Strong, and Rappel (2016)	Australia	Convenience sample of workers	Attachment, responses to sensory experiences, distress
Uljarević, Carrington, and Leekam (2016)	United Kingdom	Parents of children with Autism Spectrum Disorder	Anxiety and depression, intolerance of uncertainty
Acevedo et al. (2017)	United States	Undergraduates	Quality of childhood parenting, neural responses to emotional stimuli
Andresen et al. (2018)	Germany/ expatriates	Expatriates workers	Turnover intention, perceived stress, bonding social capital
Şengül-İnal and Sümer (2017)	Turkey	Undergraduates	
Su, Cai, and Uljarević (2018)	China	Parents of children with Autism Spectrum Disorder	Parental intolerance of uncertainty, children's comorbid symptoms, parents' mental health
Lionetti et al. (2018)	United States and United Kingdom	Undergraduates	Personality, emotional reactivity
Pazda and Thorstenson (2018)	United States	Undergraduates	Personality, color preferences, affect
Rinn, Mullet, Jett, and Nyikos (2018)		Undergraduates enrolled in a university honor program	Overexcitability
Rubaltelli et al. (2018)	Italy	Undergraduates	Risk perception and willingness to trade-off privacy, affective state, heart rate variability
Bridges and Schendan (2019)	United Kingdom	Undergraduates and community sample	Creativity
Branjerdporn, Meredith, Strong, and Green (2019)	United States	Parents	Parenting styles, adult attachment

community, worth its biological costs, only when a minority of individuals (around 20%–25%) are high on this trait, as shown in computer simulation studies (Wolf, van Doorn, & Weissing, 2008). Also, the frequency of putative susceptible genes related to an increased sensitivity to environmental influences is comparable (Belsky & Pluess, 2009).

Using a flower metaphor, highly sensitive individuals have been described as Orchids, and the rest, less sensitive population, as Dandelions (Boyce & Ellis, 2005). Orchid flowers are more sensitive to the influence of the environment, remarkably blooming in very supportive conditions (right amount of hours of light each day, right amount—and type—of water, adequate temperature, and humidity), but also promptly declining when the environment is not adequate and neglectful. On the contrary, Dandelion flowers are easy to grow, they are resilient and thrive in favorable or negative circumstances. However, this two-group classification was not explored in relation to the trait of SPS using a data-driven approach, until recently.

Two recent studies using the self-report HSP and the HSC scales across four ethnically diverse UK-based samples ages 8–19 years (total $N=3581$) and in a USA-based sample of adults ($N=906$) consistently showed that three, rather than two, SPS groups can be identified: low (25%–35%), medium (41%–47%), and high (20%–35%) (Pluess et al., 2018; Lionetti et al., 2018). This three-group categorization, applied to an independent sample of 230 UK-based adults completing the HSP scale, showed that the three sensitivity groups differed in Neuroticism, Extraversion, and emotional reactivity in response to a laboratory task designed for eliciting positive emotions. More specifically, high SPS (or Orchid) individuals scored significantly higher on Neuroticism and emotional reactivity, and lower in Extraversion, than Dandelions and the intermediate-sensitive groups. The low-SPS Dandelions also differed significantly from the intermediate-sensitive group (Lionetti et al., 2018). This intermediate-sensitive group was labeled as "Tulips," as a metaphor for a quite common flower with moderate sensitivity to the environment (not as sensitive as Orchids, Tulips require more care than Dandelion flowers, which can grow almost everywhere). Data from both UK- and USA-based samples together suggested preliminary cut-off scores of 3.8 (out of 7) between the low- and medium-sensitivity groups, and of 4.7 (out of 7) between the medium- and high-sensitive groups (considering the mean across all HSP and HSC sensitivity items). These cut-offs had relatively low sensitivity and specificity, and require replication (Lionetti et al., 2018; Pluess et al., 2018).

Currently, for the study of SPS using the sensitivity measures introduced in this chapter, the recommendation in research is to follow a two-step approach and specifically to first consider SPS as a continuous score, followed by further investigation using a group categorization approach. For group categorization, in research contexts what is recommended is to adopt a 30%-40%-30% split for low-, medium-, and high-sensitive groups based on the continuous score of the HSP and HSC scale, instead of relying on the cut-off scores, which might differ across countries.

5 Future directions

Candidate markers of SPS and of an increased sensitivity to the environment have been explored across children, adolescents, and adult samples at different levels of analysis—including genetic, neurobiological, and phenotypical. Currently, the only measures specifically developed to assess greater depth of processing in SPS are the self-report HSC for children, the HSP Scale for adults, the parent-report HSP and HSC scales, and the HSC observational rating system. These measures were described in detail and are provided in this chapter. Empirical studies and metaanalytic investigation of the degree of overlap of these measures with other temperament and personality factors have shown that these scales do not completely overlap with other existing traits. Research suggests that these scales have good psychometric proprieties. Also, intervention, correlational and laboratory studies proved that the HSP self-report scale and the HSC self-report, parent-report, and HSC rating system capture an increased sensitivity and responsivity in relation to both negative and positive stimuli, in line with theoretical models of SPS and environmental sensitivity. Although there are strengths, the reviewed self-report measures are missing two important components of SPS, as postulated at a theoretical level, which are the depth of processing and emotional reactivity. The scale factor, which most closely measures depth of processing, is the Aesthetic Sensitivity subscale. To overcome this limitation, Aron and Aron with a team of researchers with expertise on SPS are currently working to identify the items that could best capture depth of processing and emotional reactivity among the highly sensitive.

Another future direction for studies of SPS pertains to a more extensive investigation of the dimension of gender and culture. SPS theory stresses that males and females are equally likely to be highly sensitive, but still empirical data to extensively test if SPS is equally distributed across genders, including examining whether the current available measures perform in the same way irrespective of gender, are missing. For example, we can hypothesize that in Western cultures where assertiveness and boldness are promoted, particularly for men, social desirability may have an effect on sensitivity displays and being sensitive may be seen as weakness or something to disguise among males. Recent research studies adopting the HSC and HSP scales showed that SPS levels are slightly higher in females than in males, even though overall gender differences are small in effect size (for a review see Greven et al., 2019). A relatively low association between gender and SPS has been reported also in relation to the observational measure for the assessment of sensitivity in preschoolers (Lionetti, Aron, et al., 2019).

Another future direction suggested for the study of SPS involves investigation of cultural differences related to SPS as assessed with the available SPS measures. For example, currently HSP scale invariance across Belgian and British samples suggests that these specific attributes may not have a similar meaning across cultures as mean values (and thus cut-off scores) may not be completely comparable because Belgians tend to score higher on the AES scale, as also reported in Italian children (Nocentini et al., 2017; Weyn et al., 2019). Thus, cross-cultural research suggests that some of the HSP scale items might benefit from adaptations to adequately capture SPS-related cultural differences. Another consideration is that each culture specific cut-off scores based on the current SPS measure might be determined. Currently, the measure has been used in several countries, as reported in this chapter, but a more extensive analysis of the impact of culture on SPS, and on its assessment, is missing.

Another aspect of SPS that requires attention and further investigation pertains to the actual construct validity of the three SPS factors: EOE, AES, and LST, which have been consistently identified at a psychometric level. However, research has yet to clarify to what extent each of these factors contributes to an increased reactivity to environmental influences. Also, while their summary score has been predominantly used in empirical studies, their theoretical relevance and contribution to the notion of sensitivity is still unclear.

The identification of alternative tools for assessing SPS is another possible direction of study in the field of SPS and its assessment. A multiinformant, multimethod approach combining observational and parent-report and/or teacher-report measures for children, and laboratory or semistructured interviews and self-report measures for adolescents and adults, might represent useful alternatives for reaching a more comprehensive assessment of sensitivity. The use of measures other than questionnaires might be more time consuming than relying only on self-reported sensitivity, but it may lead to a more accurate evaluation of this rather complex trait.

Though not specifically developed for the assessment of SPS, other objectives markers of specific facets of sensitivity have been identified in the literature. For example, to mention a few, an adapted Visual Aesthetic Sensitivity Test (in which a subject has to express the aesthetic preference choosing between different figures) has been designed to capture sensitivity to positive stimuli, something which is closely related to aesthetic sensitivity as represented by the AES factor of SPS (Myszkowski & Storme, 2017). Facial electromyography activity (zygomaticus or "smile" muscle and the corrugator or "frown" muscle) and self-reported emotionality have been studied as markers of emotional reactivity in SPS, as well.

Interestingly, an observational measure has been recently proposed for assessing the *hawk* trait in children (Suor, Sturge-Apple, Davies, & Cicchetti, 2017). The "hawk" trait, identified in humans and animals, is thought to capture a prototypical behavioral pattern of boldness, which appears to be the opposite end of high sensitivity. Taken together, these findings suggest that other behavioral and biological mechanisms could be worthy of investigation for the identification of observable and directly measurable markers of SPS.

Finally, as introduced in this chapter, currently there are no longitudinal studies investigating the degree of stability of sensitivity across the life-span. Thanks to the recent development of SPS-informed measures for assessing sensitivity from infancy to adulthood, it is now possible to take a developmental approach to investigate SPS trajectories.

In conclusion, SPS can be assessed with a series of self-report and parent-report scales, which have been used across a number of countries, and have gone through a relatively extensive validation process. More recently, a laboratory-based observational measure for preschoolers has also been proposed, and research is moving toward the development of school-based observational and teacher-report measures. Though showing some overlap with other personality and temperament traits, research studies adopting the HSP/HSC scale and the HSC observational measure have provided evidence for SPS as a relatively distinct trait that captures individual differences in response to environmental stimuli, from early childhood to adulthood. These results have important clinical and theoretical implications, such as promoting a deeper understanding of how sensitivity develops, its early identification in children at-risk for the emergence of psychological vulnerabilities due to exposure to negative environments, and the development of an individualized prevention and intervention approach that takes into account SPS differences in response to treatment.

References

Acevedo, B. P., Aron, E. N., Aron, A., Sangster, M. D., Collins, N., & Brown, L. (2014). The highly sensitive brain: An fMRI study of sensory processing sensitivity and response to others' emotions. *Brain and Behavior, 4*(4), 580–594.

Acevedo, B. P., Jagiellowicz, J., Aron, E., Marhenke, R., & Aron, A. (2017). Sensory processing sensitivity and childhood quality's effects on neural responses to emotional stimuli. *Clinical Neuropsychiatry, 6*, 359–375.

Achenbach, T. (1991). *Manual for the child behavior checklist/4-18 and 1991 profile*. Burlington, VT: Department of Psychiatry, University of Vermont.

Ahadi, B., & Basharpoor, S. (2010). Relationship between sensory processing sensitivity, personality dimensions and mental health. *Journal of Applied Sciences, 10*(7), 570–574.

Andresen, M., Goldmann, P., & Volodina, A. (2018). Do overwhelmed expatriates intend to leave? The effects of sensory processing sensitivity, stress, and social capital on expatriates' turnover intention. *European Management Review, 15*(3), 315–328.

Aron, A., Ketay, S., Hedden, T., Aron, E. N., Rose Markus, H., & Gabrieli, J. D. (2010). Temperament trait of sensory processing sensitivity moderates cultural differences in neural response. *Social Cognitive and Affective Neuroscience, 5*(2–3), 219–226.

Aron, E. (1996). *The highly sensitive person: How to thrive when the worldoverwhelms you*. New York: Broadway Books.

Aron, E. N. (2002). *The highly sensitive child: Helping our children thrive when the world overwhelms them*. New York: Broadway Books.

Aron, E. N., & Aron, A. (1997). Sensory-processing sensitivity and its relation to introversion and emotionality. *Journal of Personality and Social Psychology, 73*(2), 345–368.

Aron, E. N., Aron, A., & Davies, K. M. (2005). Adult shyness: The interaction of temperamental sensitivity and an adverse childhood environment. *Personality and Social Psychology Bulletin, 31*(2), 181–197.

Aron, E. N., Aron, A., & Jagiellowicz, J. (2012). Sensory processing sensitivity: A review in the light of the evolution of biological responsivity. *Personality and Social Psychology Review, 16*(3), 262–282.

Aron, E. N., Aron, A., & Tillmann, T. (2018). *Future directions for sensory processing sensitivity research*. In *Presentation as part of the symposium "Novel Perspectives on Sensory Processing Sensitivity" at the 30th convention of the Association for Psychological Science (APS) in San Francisco, California, May 25, 2018*.

Bakker, K., & Moulding, R. (2012). Sensory-processing sensitivity, dispositional mindfulness and negative psychological symptoms. *Personality and Individual Differences, 53*(3), 341–346.

Belsky, J., Bakermans-Kranenburg, M. J., & Van IJzendoorn, M. H. (2007). For better and for worse: Differential susceptibility to environmental influences. *Current Directions in Psychological Science, 16*(6), 300–304.

Belsky, J., & Pluess, M. (2009). Beyond diathesis stress: Differential susceptibility to environmental influences. *Psychological Bulletin, 135*(6), 885–908.

Benham, G. (2006). The highly sensitive person: Stress and physical symptom reports. *Personality and Individual Differences, 40*(7), 1433–1440.

Booth, C., Standage, H., & Fox, E. (2015). Sensory-processing sensitivity moderates the association between childhood experiences and adult life satisfaction. *Personality and Individual Differences, 87*, 24–29.

Boterberg, S., & Warreyn, P. (2016). Making sense of it all: The impact of sensory processing sensitivity on daily functioning of children. *Personality and Individual Differences, 92*, 80–86.

Boyce, W. T., & Ellis, B. J. (2005). Biological sensitivity to context: I. An evolutionary-developmental theory of the origins and functions of stress reactivity. *Development and Psychopathology, 17*(2), 271–301.

Branjerdporn, G., Meredith, P., Strong, J., & Green, M. (2019). Sensory sensitivity and its relationship with adult attachment and parenting styles. *PLoS ONE, 14*(1), e0209555.

Bridges, D., & Schendan, H. E. (2019). The sensitive, open creator. *Personality and Individual Differences, 142*(1), 179–185.

Brindle, K., Moulding, R., Bakker, K., & Nedeljkovic, M. (2015). Is the relationship between sensory-processing sensitivity and negative affect mediated by emotional regulation? *Australian Journal of Psychology, 67*(4), 214–221.

Chen, C., Chen, C., Moyzis, R., Stern, H., He, Q., Li, H., ... Dong, Q. (2011). Contributions of dopamine-related genes and environmental factors to highly sensitive personality: A multi-step neuronal system-level approach. *PLoS ONE, 6*(7), e21636.

Chen, C., Xiu, D., Chen, C., Moyzis, R., Xia, M., He, Y., ... Wang, Y. (2015). Regional homogeneity of resting-state brain activity suppresses the effect of dopamine-related genes on sensory processing sensitivity. *PLoS ONE, 10*(8), e0133143.

Dyson, M. W., Olino, T. M., Dunn, W., Goldsmith, H. H., & Klein, D. N. (2012). The structure of temperament in preschoolers: A two-stage factor analytic approach. *Emotion, 12*(1), 44–57.

Ellis, B. J., & Boyce, W. T. (2008). Biological sensitivity to context. *Current Directions in Psychological Science, 17*(3), 183–187.

Evans, D. E., & Rothbart, M. K. (2008). Temperamental sensitivity: Two constructs or one? *Personality and Individual Differences, 44*(1), 108–118.

Evers, A., Rasche, J., & Schabracq, M. J. (2008). High sensory-processing sensitivity at work. *International Journal of Stress Management, 15*(2), 189–198.

Gearhart, C. C. (2014). Sensory-processing sensitivity and nonverbal decoding: The effect on listening ability and accuracy. *International Journal of Listening, 28*(2), 98–111.

Gearhart, C. C., & Bodie, G. D. (2012). Sensory-processing sensitivity and communication apprehension: Dual influences on self-reported stress in a college student sample. *Communication Reports, 25*(1), 27–39.

Gerstenberg, F. X. (2012). Sensory-processing sensitivity predicts performance on a visual search task followed by an increase in perceived stress. *Personality and Individual Differences, 53*(4), 496–500.

Ghorbani Taghlidabad, B., & Tasbihsazan Mashhadi, R. (2018). Investigating the relationship between sensory processing sensitivity and life style with stress in patients with irritable bowel syndrome. *Journal of Practice in Clinical Psychology, 6*(4), 239–248.

Goldberg, A., Ebraheem, Z., Freiberg, C., Ferarro, R., Chai, S., & Gottfried, O. D. (2018). Sweet and sensitive: Sensory processing sensitivity and type 1 diabetes. *Journal of Pediatric Nursing, 38*, e35–e38.

Goodman, R. (1997). The strengths and difficulties questionnaire: A research note. *Journal of Child Psychology and Psychiatry, 38*, 581–586.

Gray, J. A. (1981). A critique of Eysenck's theory of personality. In H. J. Eysenck (Ed.), *A model for personality* (pp. 246–276). Berlin: Springer-Verlag.

Gray, J. A. (1982). *The neuropsychology of anxiety: An enquiry into thefunctions of the septo-hippocampal system.* New York: Oxford University Press.

Greven, C. U., Lionetti, F., Booth, C., Aron, E., Fox, E., Schendan, H. E., ... Bijttebier, P. (2019). Sensory Processing Sensitivity in the context of environmental sensitivity: A critical review and development of research agenda. *Neuroscience and Biobehavioral Reviews, 98*, 287–305.

Hofmann, S. G., & Bitran, S. (2007). Sensory-processing sensitivity in social anxiety disorder: Relationship to harm avoidance and diagnostic subtypes. *Journal of Anxiety Disorders, 21*(7), 944–954.

Jagiellowicz, J., Aron, A., & Aron, E. N. (2016). Relationship between the temperament trait of sensory processing sensitivity and emotional reactivity. *Social Behavior and Personality: An International Journal, 44*(2), 185–199.

Jagiellowicz, J., Xu, X., Aron, A., Aron, E. N., Cao, G., Feng, T., & Weng, X. (2012). The trait of sensory processing sensitivity and neural responses to changes in visual scenes. *Social Cognitive and Affective Neuroscience*, *6*(1), 38–47.

Jonsson, K., Grim, K., & Kjellgren, A. (2014). Do highly sensitive persons experience more nonordinary states of consciousness during sensory isolation? *Social Behavior and Personality: An Internationa Journal*, *42*(9), 1495–1506.

Kagan, J. (1989). Temperamental contributions to social behavior. *American Psychologist*, *44*(4), 668.

Kagan, J., Reznick, J. S., & Snidman, N. (1987). The physiology and psychology of behavioral inhibition in children. *Child Development*, *58*(6), 1459–1473.

Karreman, A., Van Tuijl, C., van Aken, M. A., & Deković, M. (2006). Parenting and self-regulation in preschoolers: A meta-analysis. *Infant and Child Development*, *15*(6), 561–579.

Kibe, C., Suzuki, M., & Hirano, M. (2018). *Individual sensitivity to the effects of resilience education: Self-esteem enhancement in Japanese adolescents*. In *Paper presented at the European conference on positive psychology, Budapest, Hungary*.

Kim, S., & Kochanska, G. (2012). Child temperament moderates effects of parent–child mutuality on self-regulation: A relationship-based path for emotionally negative infants. *Child Development*, *83*, 1275–1289.

Kjellgren, A., Lindahl, A., & Norlander, T. (2009). Altered states of consciousness and mystical experiences during sensory isolation in flotation tank: Is the highly sensitive personality variable of importance? *Imagination, Cognition and Personality*, *29*(2), 135–146.

Konrad, S., & Herzberg, P. Y. (2017). Psychometric properties and validation of a German High Sensitive Person Scale (HSPS-G). *European Journal of Psychological Assessment*, *35*(3), 364–378.

Lionetti, F., Aron, A., Aron, E. N., Burns, L. G., Jagiellowicz, J., & Pluess, M. (2018). Dandelions, Tulips and Orchids: Evidence for the existence of low-sensitive, medium-sensitive, and high-sensitive individuals. *Translational Psychiatry*, *8*(24), 6.

Lionetti, F., Aron, E. N., Aron, A., Klein, D. N., & Pluess, M. (2019). Observer-rated environmental sensitivity moderates children's response to parenting quality in early childhood. *Developmental Psychology*, https://doi.org/10.1037/dev0000795.

Lionetti, F., Pastore, M., Moscardino, U., Nocentini, A., Pluess, K., & Pluess, M. (2019). Sensory Processing Sensitivity and its association with personality traits and affect: A meta-analysis. *Journal of Research in Personality*, *81*, 138–152.

Liss, M., Mailloux, J., & Erchull, M. J. (2008). The relationships between sensory processing sensitivity, alexithymia, autism, depression, and anxiety. *Personality and Individual Differences*, *45*(3), 255–259.

Liss, M., Timmel, L., Baxley, K., & Killingsworth, P. (2005). Sensory processing sensitivity and its relation to parental bonding, anxiety, and depression. *Personality and Individual Differences*, *39*(8), 1429–1439.

Listou Grimen, H., & Diseth, Å. (2016). Sensory Processing Sensitivity: Factors of the highly sensitive person scale and their relationships to personality and subjective health complaints. *Perceptual and Motor Skills*, *123*(3), 637–653.

May, A. (2019). Sensory Processing Sensitivity in a South Africa community sample (in preparation).

McNaughton, N., & Gray, J. A. (2000). Anxiolytic action on the behavioural inhibition system implies multiple types of arousal contribute to anxiety. *Journal of Affective Disorders*, *61*(3), 161–176.

Meredith, P. J., Bailey, K. J., Strong, J., & Rappel, G. (2016). Adult attachment, sensory processing, and distress in healthy adults. *American Journal of Occupational Therapy*, *70*(1), 1–8.

Meyer, B., Ajchenbrenner, M., & Bowles, D. P. (2005). Sensory sensitivity, attachment experiences, and rejection responses among adults with borderline and avoidant features. *Journal of Personality Disorders*, *19*(6), 641–658.

Meyer, B., & Carver, C. S. (2000). Negative childhood accounts, sensitivity, and pessimism: A study of avoidant personality disorder features in college students. *Journal of Personality Disorders, 14*(3), 233–248.

Monroe, S. M., & Simons, A. D. (1991). Diathesis-stress theories in the context of life stress research: Implications for depressive disorders. *Psychological Bulletin, 110*, 406–425.

Myszkowski, N., & Storme, M. (2017). Measuring "Good Taste" with the Visual Aesthetic Sensitivity Test-Revised (VAST-R). *Personality and Individual Differences, 117*, 91–100.

Neal, J. A., Edelmann, R. J., & Glachan, M. (2002). Behavioural inhibition and symptoms of anxiety and depression: Is there a specific relationship with social phobia? *British Journal of Clinical Psychology, 41*(4), 361–374.

Nocentini, A., Menesini, E., Lionetti, F., & Pluess, M. (2017). *Highly Sensitive Child scale: Investigation of construct validity in a sample of Italian children.* In *Paper presented at the European conference of developmental psychology, Utrecht.*

Nocentini, A., Menesini, E., & Pluess, M. (2018). The personality trait of environmental sensitivity predicts children's positive response to school-based anti-bullying intervention. *Clinical Psychological Science, 6*(6), 848–859.

Palladino, B. E., Nocentini, A., & Menesini, E. (2015). Psychometric properties of the florence cyberbullying-cybervictimization scales. *Cyberpsychology, Behavior, and Social Networking, 18*(2), 112–119.

Pazda, A. D., & Thorstenson, C. A. (2018). Extraversion predicts a preference for high-chroma colors. *Personality and Individual Differences, 127*, 133–138.

Pluess, M. (2015). Individual differences in environmental sensitivity. *Child Development Perspectives, 9*(3), 138–143.

Pluess, M., Assary, E., Lionetti, F., Lester, K. J., Krapohl, E., Aron, E. N., & Aron, A. (2018). Environmental sensitivity in children: Development of the highly sensitive child scale and identification of sensitivity groups. *Developmental Psychology, 54*(1), 51–70.

Pluess, M., & Belsky, J. (2010). Differential susceptibility to parenting and quality child care. *Developmental Psychology, 46*(2), 379–390.

Pluess, M., & Belsky, J. (2013). Vantage sensitivity: Individual differences in response to positive experiences. *Psychological Bulletin, 139*(4), 901–916.

Pluess, M., & Boniwell, I. (2015). Sensory-Processing Sensitivity predicts treatment response to a school-based depression prevention program: Evidence of Vantage Sensitivity. *Personality and Individual Differences, 82*, 40–45.

Pluess, M., & Lionetti, F. (2018). *The Highly Sensitive Child scale—Teacher report.* [Unpublished manuscript].

Pluess, M., Lionetti, F., Aron, E., & Aron, A. (2020). People differ in their sensitivity to the environment: Measurement, association with personality traits and experimental evidence (submitted).

Radloff, L. S. (1977). The CES-D scale: A self-report depression scale for research in the general population. *Applied Psychological Measurement, 1*(3), 385–401.

Reise, S. P., Bonifay, W. E., & Haviland, M. G. (2013). Scoring and modeling psychological measures in the presence of multidimensionality. *Journal of Personality Assessment, 95*(2), 129–140.

Reise, S. P., Moore, T. M., & Haviland, M. G. (2010). Bifactor models and rotations: Exploring the extent to which multidimensional data yield univocal scale scores. *Journal of Personality Assessment, 92*(6), 544–559, https://doi.org/10.1080/00223891.2010.496477.

Riley, A. W., Forrest, C. B., Rebok, G. W., Starfield, B., Green, B. F., Robertson, J. A., & Friello, P. (2004). The child report form of the CHIP-child edition: Reliability and validity. *Medical Care*, 221–231.

Riley, A. W., Forrest, C. B., Starfield, B., Rebok, G. W., Robertson, J. A., & Green, B. F. (2004). The parent report form of the CHIP-Child Edition: Reliability and validity. *Medical Care*, 210–220.

Rinn, A. N., Mullet, D. R., Jett, N., & Nyikos, T. (2018). Sensory processing sensitivity among high-ability individuals: A psychometric evaluation of the highly sensitive person scale. *Roeper Review, 40*(3), 166–175.

Rothbart, M. K., & Bates, J. E. (2006). Temperament. In N. Eisenberg, W. Damon, & R. M. Lerner (Eds.), *Handbook of child psychology: Vol. 3, social, emotional, and personality development* (6th ed., pp. 99–166). Hoboken, NJ: John Wiley & Sons Inc.

Rubaltelli, E., Scrimin, S., Moscardino, U., Priolo, G., & Buodo, G. (2018). Media exposure to terrorism and people's risk perception: The role of environmental sensitivity and psychophysiological response to stress. *British Journal of Psychology, 109*(4), 656–673.

Rutter, M., & Sroufe, L. A. (2000). Developmental psychopathology: Concepts and challenges. *Development and psychopathology, 12*(3), 265–296.

Scrimin, S., Osler, G., Pozzoli, T., & Moscardino, U. (2018). Early adversities, family support, and child well-being: The moderating role of environmental sensitivity. *Child: Care, Health and Development, 44*(6), 885–891.

Şengül-İnal, G., & Sümer, N. (2017). Exploring the multidimensional structure of sensory processing sensitivity in Turkish samples. *Current Psychology*, 1–13, https://doi.org/10.1007/s12144-017-9751-0.

Silvestre, A. (2018). *Actuaciones específicas en el aula para alumnos con necesidades educativas especiales: el rasgo de la alta sensibilidad* [Master the Profesorado]. Universidad de Zaragoza [Unpublished Manuscript].

Shiota, M. N., Keltner, D., & Mossman, A. (2007). The nature of awe: Elicitors, appraisals, and effects on self-concept. *Cognition and Emotion, 21*(5), 944–963.

Slagt, M., Dubas, J. S., Dekovic, M., & van Aken, M. A. (2016). Differences in sensitivity to parenting depending on child temperament: A meta-analysis. *Psychological Bulletin, 142*(10), 1068.

Slagt, M., Dubas, J. S., van Aken, M. A., Ellis, B. J., & Dekovic, M. (2018). Sensory processing sensitivity as a marker of differential susceptibility to parenting. *Developmental Psychology, 54*(3), 543–558.

Smolewska, K. A., McCabe, S. B., & Woody, E. Z. (2006). A psychometric evaluation of the Highly Sensitive Person Scale: The components of sensory-processing sensitivity and their relation to the BIS/BAS and "Big Five". *Personality and Individual Differences, 40*(6), 1269–1279.

Sobocko, K., & Zelenski, J. M. (2015). Trait sensory-processing sensitivity and subjective well-being: Distinctive associations for different aspects of sensitivity. *Personality and Individual Differences, 83*, 44–49.

Suor, J. H., Sturge-Apple, M. L., Davies, P. T., & Cicchetti, D. (2017). A life history approach to delineating how harsh environments and hawk temperament traits differentially shape children's problem-solving skills. *Journal of Child Psychology and Psychiatry, 58*(8), 902–909.

Su, X., Cai, R. Y., & Uljarević, M. (2018). Predictors of mental health in Chinese parents of children with autism Spectrum disorder (ASD). *Journal of Autism and Developmental Disorders, 48*(4), 1159–1168.

Þórarinsdóttir, Þ.K. (2018). *Psychometric properties of the highly sensitive person scale and its relationship to the big five personality traits in a sample of Icelandic University students* [Doctoral dissertation]. University of Reykjavik.

Tillmann, T., El Matany, K., & Duttweiler, H. (2018). Measuring environmental sensitivity in educational contexts: A validation study with German-speaking students. *Journal of Educational and Developmental Psychology, 8*, 17–28.

Uljarević, M., Carrington, S., & Leekam, S. (2016). Brief report: Effects of sensory sensitivity and intolerance of uncertainty on anxiety in mothers of children with autism spectrum disorder. *Journal of Autism and Developmental Disorders, 46*(1), 315–319.

Wachs, T. D. (2013). Relation of maternal personality to perceptions of environmental chaos in the home. *Journal of Environmental Psychology*, *34*, 1–9.
Weyn, S., Van Leeuwen, K., Pluess, M., Lionetti, F., Greven, C., Goosens, L., ... Bijttebier, P. (2019). Psychometric properties of the highly sensitive child scale across developmental stage, gender, and country. *Current Psychology*, https://doi.org/10.1007/s12144-019-00254-5.
Wolf, M., van Doorn, G. S., & Weissing, F. J. (2008). Evolutionary emergence of responsive and unresponsive personalities. *Proceedings of the National Academy of Sciences of the United States of America*, *105*, 15825–15830.
Woods, R. L., Colvin, C. R., Vera-Diaz, F. A., & Peli, E. (2010). A relationship between tolerance of blur and personality. *Investigative Ophthalmology & Visual Science*, *51*(11), 6077–6082.
Woodward, S. A., Lenzenweger, M. F., Kagan, J., Snidman, N., & Arcus, D. (2000). Taxonic structure of infant reactivity: Evidence from a taxometric perspective. *Psycholological Science*, *11*, 296–301.

CHAPTER 3

Sensory processing sensitivity—For better or for worse? Theory, evidence, and societal implications

Corina U. Greven[a,b,c], **Judith R. Homberg**[a]
[a]Department of Cognitive Neuroscience, Donders Institute for Brain, Cognition and Behaviour, Radboud University Medical Center, Nijmegen, The Netherlands
[b]Karakter Child and Adolescent Psychiatry University Center, Nijmegen, The Netherlands
[c]Social, Genetic & Developmental Psychiatry Centre, Institute of Psychiatry, Psychology & Neuroscience, King's College London, London, United Kingdom

Contents

1 SPS in the context of theories on environmental sensitivity 52
1.1 Diathesis-stress theory 52
1.2 Biological sensitivity to context theory 53
1.3 Differential susceptibility theory 54
1.4 Vantage sensitivity theory 55
1.5 Sensory processing sensitivity theory 55
1.6 A metaframework of environmental sensitivity 56
1.7 Other considerations 59
1.8 Implications for research on SPS 60
2 Empirical evidence for SPS as a marker for environmental sensitivity 63
2.1 Evidence favoring diathesis-stress 63
2.2 Evidence favoring differential susceptibility and vantage sensitivity 65
2.3 Other evidence for SPS as a marker of environmental sensitivity 67
2.4 Conclusion 67
3 Implications and future directions 68
4 Conclusion 70
References 71

Humans and other animal species are sensitive to the environment. However, some individuals are substantially more sensitive to the effects of the environment compared to others. Sensory processing sensitivity (SPS), the scientific term for high sensitive personality (HSP), is a biological and psychological trait associated with heightened sensitivity to the environment. The term "environment" is here broadly defined, referring to salient

https://doi.org/10.1016/B978-0-12-818251-2.00003-5

internal or external stimuli, including physical environments (e.g., caffeine intake, food), social environments (e.g., experiences in childhood, crowds, other's moods), sensory stimuli (e.g., visual, auditory, tactile, olfactory), and internal events (e.g., thoughts, feelings, bodily sensations such as hunger, pain) (Greven et al., 2019). It also includes conditioned "neutral" stimuli that predict salient unconditioned stimuli.

Who should read this chapter? This chapter is relevant to anyone interested in scientific theories on SPS and environmental sensitivity, as well as their practical applications. In this chapter, we will start by outlining major theories of what makes some individuals more sensitive to the environment than others. There are two sides to the coin of being highly sensitive. It means being more affected by negative environments, as well as positive ones. Therefore, the second part of this chapter integrates empirical evidence showing that higher levels of SPS make a person more sensitive to positive and/or negative environments. The last part of this chapter is focused on discussing broader and practical implications of high sensitivity for society, including general public health, employment, and education sectors. In our discussion of SPS, we focus on peer-reviewed scientific articles published in indexed journals.

1 SPS in the context of theories on environmental sensitivity

1.1 Diathesis-stress theory

The diathesis-stress model (Gottesman & Shields, 1967), also called the dual-risk or transactional model (Sameroff, 1983), posits that some individuals are more vulnerable to developing disorders and illness when faced with environmental stressors as a result of underlying vulnerability factors. These factors can, for instance, be behavioral (e.g., "difficult temperament"), physiological (e.g., heightened stress system reactivity), or genetic (e.g., short allele of the serotonin transporter-linked polymorphic region 5-HTTLPR) (Ellis, Boyce, Belsky, Bakermans-Kranenburg, & van IJzendoorn, 2011). The term dual-risk refers to the interaction between the risk inherent factor carried by a person on the one hand (e.g., genetic factors, temperament traits), and the environmental stressor on the other hand. Individuals without these vulnerability characteristics are considered resilient, that is, not affected by environmental stressors. The model does not explicitly address responsivity to positive, supportive environmental conditions. Yet, implicit in this model is the assumption that vulnerable and resilient individuals respond similarly to positive, supportive environments. Hence, the difference between these two groups of individuals becomes apparent in situations of adversity. Thus, diathesis-stress is a deficit-focused biomedical model of sensitivity to the environment that provides a grounded approach for understanding individual differences to adversity.

1.2 Biological sensitivity to context theory

Biological sensitivity to context (BSC) theory started with empirical evidence on cardiovascular and immune reactivity in response to environmental adversity in 3- to 5-year-old children (Boyce et al., 1995). The research termed these very sensitive children as "highly reactive," showing the highest illness incidences in case of high adversity, consistent with diathesis-stress theory. However, the unexpected finding with BSC, not consistent with diathesis-stress, was that the "highly reactive" children, when living in a supportive family setting, showed the lowest illness rates (Boyce et al., 1995).

This work provided the basis for BSC theory, which posits that the same response systems that increase vulnerability to problematic psychiatric and biomedical outcomes in situations of environmental adversity, also facilitate greater mental and physical well-being in supportive and resource-abundant environments (Ellis et al., 2011). Such susceptibility to both negative *and* positive environments, based on the same underlying factors, is now referred to as differential susceptibility (differential susceptibility (DS) theory, outlined in Section 1.3). As discussed in Chapter 2, in an analogy to a Swedish idiom, BSC theory referred to highly reactive children as "orchids," who are sensitive to positive and negative environments in a for "better-and-for worse" fashion. Less reactive children were referred to as "dandelions," who are less responsive to environmental stimulation (Boyce & Ellis, 2005). In Swedish "orkidebarn" stands for "orchid child," as orchids typically show no flowers in poor environmental conditions, but flourish and give flowers considered by many of outstanding beauty when they are exposed to environmental conditions suited to them. "Maskrosbarn" stands for dandelion child, as dandelions grow well at different levels of typical environmental conditions. BSC theory emphasizes physiological mechanisms (such as autonomic, adrenocortical, and immune reactivity to psychosocial stressors) as the main basis for heightened responsivity to the environment. BSC theory, rooted in evolutionary theory, proposes conditional adaptation: that is, it suggests that environments early in life shape and program how an individual adapts to future environmental conditions. In particular, BSC theory suggests a U-shaped relationship between early environmental adversity or support, and psychosocial outcomes. This U-shape reflects the hypothesis that highly reactive children ("orchids") arise from the most supportive, as well as most stressful, childhood environments. The majority of children ("dandelions") raised in typical environments have lower levels of biological sensitivity to context. This allows them to function in a world that is neither particularly threatening nor safe (Boyce & Ellis, 2005). Genes and gene-environment interactions also play a role in shaping biological sensitivity to context. Genes, according to BSC theory, set the "reaction norm," that is, they determine the range in which one can adapt to the environment (Boyce & Ellis, 2005). For example, a child with a genetic predisposition to being environmentally sensitive, will be more likely to become highly sensitive when exposed to very supportive or stressful childhood environments, relative to a child that does not have such a genetic predisposition.

1.3 Differential susceptibility theory

Differential susceptibility (DS) theory came from a purely theoretical background, in response to critiques of Belsky's evolutionary theory of socialization (Belsky, 2000). Although DS theory arose independently from BSC theory, the theories have much in common. Like BSC theory, DS theory posits that certain characteristics make some individuals disproportionately more susceptible to negative, risk-promoting environments, but also make them more sensitive to positive, development-enhancing environmental conditions (differential susceptibility as a mechanism) (Belsky & Pluess, 2009). DS theory is also grounded in evolutionary thinking, but in contrast to BSC theory, it emphasizes parental bet-hedging (Belsky & Pluess, 2009). According to Belsky and Pluess (2009), children adapt to early life experiences to prepare them for future challenges and opportunities—known as conditional adaptation. Yet, because the future is uncertain and may turn out different from what was expected, not one single strategy will be optimal. Hence, according to this thinking, parents produce offspring of varying developmental strategies: some "conditional" have greater plasticity and sensitivity, and others "alternative" have more stable, fixed systems. These strategies are thought to be maintained by natural selection. Conditional strategies are shaped by early environmental factors to prepare a good fit to the environment, whereas alternative strategies are less influenced by environmental conditions and more fixed. These variable strategies allow parents to produce offspring that are guarded against potential misfit to future environmental conditions given uncertainty about the future (alternative), as well, ensuring other offspring that may be particularly well prepared for future environmental conditions (conditional). Hence, children within a family should differ in developmental strategies, thereby ensuring reproductive fitness and survival of the family genes. Interindividual differences in differential susceptibility are thought to be influenced by genetics, neurobiology, as well as prenatal and early postnatal factors (Pluess & Belsky, 2011).

1.4 Vantage sensitivity theory

Vantage sensitivity (VS) theory emerged from the observation that much attention had been directed to interindividual differences in responsivity to negative, but not positive environments, possibly due to an absence of theoretical frameworks (Pluess & Belsky, 2013). VS theory posits that certain characteristics make some people disproportionally more susceptible to positive experiences (Pluess & Belsky, 2013). Vantage sensitivity may be seen as a short form for advantage, but also refers to a position or place affording advantage or superiority (Pluess & Belsky, 2013). It is not the same as "protection," "buffering," or "resilience," which refers to the result of not being adversely affected by a negative experience. The term vantage resistance describes the inability to benefit from environmental support and enrichment. VS theory is derived from and closely related to DS theory, yet it differs from DS theory as it reflects more than just "the positive side" of differential susceptibility. This is because individuals showing vantage sensitivity, may at the same time not be vulnerable to, or even resilient to adversity. One example given in the literature is IQ, which may help children to benefit proportionally more from a high-quality education, but does not mean that a child with high IQ would proportionally be more adversely affected by a low-quality education (Pluess & Belsky, 2013). In sum, VS theory is the first theoretical framework to draw attention to interindividual differences in responsivity to positive environments.

1.5 Sensory processing sensitivity theory

As SPS theory is thoroughly described throughout this book, the focus here is on summarizing important differences between SPS theory and the aforementioned theories. SPS theory was developed based on extensive literature on temperament and behavioral response to stimuli in humans and animals (Aron & Aron, 1997). In addition to this body of work, SPS theory was derived from in-depth qualitative interviews conducted with adults who self-identified as being highly sensitive, introverted, or easily overwhelmed by stimuli (Aron & Aron, 1997). SPS is considered to be an evolutionarily conserved trait observed in >100 species, reflecting a continuum of interindividual differences in depth of processing and emotional reactivity to environmental information (Acevedo et al., 2014; Acevedo, Jagiellowicz, Aron, Aron, & Marhenke, 2017; Greven et al., 2019). SPS is therefore a fundamental factor in understanding how behavior is guided by environments, negative and positive ones, and is also hypothesized to reflect differential susceptibility. SPS theory also highlights the neurobiological and genetic underpinnings of the SPS trait (Aron & Aron, 1997; Aron, Aron, & Jagiellowicz, 2012). High SPS is proposed to be present in a minority of the population (Aron & Aron, 1997; Aron et al., 2012). BSC theory and DS theory also share much in common with SPS theory. What differs is that BSC theory and DS theory emerged from childhood developmental perspectives (Ellis et al., 2011), whereas SPS theory was largely developed from perspectives and research on interindividual differences in cognitive processes and emotional reactivity in adults (Aron & Aron, 1997). Unique to SPS theory is that it is the first theory, which summarizes sensitivity to the environment in a personality trait reflecting depth-of-processing, emotional reactivity and empathy, sensitivity to subtleties, and ease of overstimulation. Also, several psychometric advances have permitted assessment of SPS, captured with the Highly Sensitive Person (HSP) scale for adults (Aron & Aron, 1997). The HSP scale was later converted to a version for children: the Highly Sensitive Child (HSC) Scale (Pluess et al., 2017). Since the development of the HSP scale, several scales have been developed, mainly for assessment of high-sensitivity in children as detailed in Chapter 2. Also, subsequent studies using these scales have elaborated on the neural and genetic basis of SPS in humans (for review see; Acevedo, Aron, Pospos, & Jessen, 2018) and animals (Homberg, Schubert, Asan, & Aron, 2016), as discussed in Chapter 5.

1.6 A metaframework of environmental sensitivity

It is interesting that multiple theoretical frameworks on sensitivity to the environment emerged simultaneously, in parallel, but independent from each other. This has been linked to the idea of "multiple discoveries" (Merton, 1973). "Discoveries are often made concurrently, by multiple investigators, when a line of thinking or experimentation 'ripens' into an accessible and novel theoretical account" (Boyce, 2016). Each of the mentioned theories provides a unique perspective on interindividual differences in sensitivity and responsivity to the environment. A key feature common across BSC, DS theory, and SPS theory is that they all emphasize differential susceptibility: sensitivity or responsivity not only to environmental adversity, but also to positive features of the environment. Diathesis-stress and VS theories focus on responsivity to exclusively negative and positive environments, respectively. Understanding differences and similarities between the various theories can seem challenging, as the theories use different terminology, and due to different backgrounds of the authors developing the theories. Yet, integration of theories is useful to gain further insights.

Such an integrated perspective was more recently achieved in the metaframework of environmental sensitivity (ES) (Pluess, 2015). From the ES metaframework, the formation of the Neurosensitivity hypothesis arose (Pluess, 2015). This hypothesis suggests that the sensitivity of the central nervous system (CNS) is shaped by early environments, genetic markers indexing sensitivity, and the interaction of the two. Sensitivity of the CNS is in turn expressed in terms of psychological and physiological responsivity. Integrating evolutionary thinking, the ES metaframework suggests that conditional adaptation and bet-hedging strategies are not exclusive, but that a combination of these strategies may be taking place. Further, computer simulation and empirical evidence suggest that high sensitivity only holds evolutionary advantage if a minority of the population is sensitive, typically around 20% (Wolf, van Doorn, & Weissing, 2008). In fact, BSC, DS theory, and SPS theory all converge on the proposition that only a minority of the population is highly sensitive.

The ES metaframework also makes a differentiation between sensitivity types (Pluess, 2015). Here a number of speculations about the etiologies of these sensitivity types are made, which remain to be tested thoroughly. In the absence of genetic variants associated with high sensitivity, environmental sensitivity is assumed to be low, independent of environmental exposure. However, if genetic variants associated with sensitivity are present, the quality of the environment is thought to impact the formation of sensitivity type: if the environment is generally neutral or normative, this will lead to the development of general sensitivity to both positive and negative environments (differential susceptibility). If the environment is mainly adverse, this is thought to lead to the development of increased sensitivity toward threats (diathesis-stress). If the environment is predominantly supportive, then this would result in sensitivity toward supportive environments and enrichment (vantage sensitivity). In practice, a large number of genes may modulate sensitivity (also referred to as plasticity genes), and although some have been identified, many remain to be discovered. However, people are most likely to differ in how many variants of such plasticity genes they carry, rather than there being individuals with and without a particular "sensitivity gene." Through this lens, sensitivity may be thought of as multifaceted and highly flexible, being both affected by variation in genes and life experiences, including childhood environments.

Furthermore, the ES framework differentiates conceptually between two different aspects of environmental sensitivity: developmental plasticity and immediate reactivity (Pluess, 2015). Developmental plasticity refers to changes that arise early in development, in a limited timeframe, and allow adaptations to the environment through developmental programming. However, according to developmental plasticity, additional adaptations in response to changes later in life are unlikely. In contrast, immediate reactivity refers to sensitivity as a relatively stable trait in response to varying environmental contexts, not only in childhood, but across the lifespan and in different situations.

What is more, the ES framework also makes a differentiation between sensitivity and reactivity (Pluess, 2015). Whereas "sensitivity" refers to the underlying propensity to perceive and internally process external influences, "responsivity" refers to the behavioral consequence of sensitivity. However, these two terms are often used interchangeably in the literature, including in this chapter.

Lastly, an important conclusion is that all theories integrated in the ES framework emphasize the substantial interindividual variation in how different species show sensitivity to environments. In many lines of scientific investigation, such variation is considered to be "error" or "noise," and therefore ignored, instead of basing analyses and conclusions on studying mean differences between groups. For example, studying the influence of intervention A (i.e., group A) on internalizing problems, relative to intervention B (i.e., group B), both interventions may turn out to be equally effective (i.e., no differences in means). Yet, the focus on mean differences ignores that for both interventions (A and B), there may have been some individuals who responded well to the intervention, and others who did not. Studies failing to take interindividual variation into account tend to underestimate effect sizes. This is because effects may be stronger in more sensitive individuals, but weaker or absent in less sensitive individuals. Theories of ES strongly argue for the importance of taking interindividual differences in environmental sensitivity into account across all lines of scientific investigation. These perspectives suggest a more personalized approach, whether the goal is to prevent or ameliorate adverse outcomes, or foster well-being and positive functioning. As Boyce (2016) put it, "inter-individual variation is not 'noise', but 'music' to which our best scientific ears should be attuned."

1.7 Other considerations

In addition to SPS theory, there are also other temperament and personality traits moderating sensitivity to the environment, such as introversion, high reactive temperament, openness to experience, and behavioral inhibition (Aron et al., 2012; Boyce & Ellis, 2005; Pluess, 2015). Although these traits and theories differ from each other, they all have in common that they indicate how individuals vary in response to the environment. Therefore, a reasonable question arises whether the construct of SPS may already adequately be captured by other temperament or personality traits. However, evidence across multiple samples suggests that existing personality and temperament traits, when taken together can, at best, explain a modest amount of variance (around 35%) related to SPS (Pluess et al., 2017). Further, SPS may reflect a specific, not previously described constellation of facets of the Big Five personality traits, which are thought to reflect basic dimensions of personality (Greven et al., 2019). However, more research clarifying the relationship between SPS with other temperament and personality traits is needed, such as further research on incremental validity beyond the Big Five personality dimensions.

The mechanisms underlying differential susceptibility to the environment are slowly beginning to be unraveled. These include children's temperament and behaviors, as well as differences in autonomic and adrenocortical reactivity, together with mechanisms at the levels of brain circuitry, synapse and cell, plasticity genes and epigenome (for a review see Boyce, 2016). A mechanism that has not received so much emphasis yet is how different styles of cognitive processing emerge. This topic has been put forth by the Cognitive Bias (CogBIAS) hypothesis (Fox & Beevers, 2016), which suggests that sensitivity genes may interact with environments to influence the development of "toxic" or "enhancing" cognitive biases, which in turn affect the development of problem behavior or positive functioning. Toxic cognitive biases could emerge from early adverse environments, and are thought to bias future information-processing that reinforces sensitivity to negative environments, thereby increasing risk for psychopathology. In contrast, a supportive early environment would promote "enhancing" cognitive biases, such as selective attention for positive information, thus increasing the chances that an individual will thrive and show high levels of functioning and well-being. Importantly, cognitive biases can be modified through intervention. For example, a negative attentional bias modification training has been shown to increase stress reactivity in adult participants (MacLeod, Rutherford, Campbell, Ebsworthy, & Holker, 2002). In contrast, interventions that target increasing attention to positive stimuli have been shown to reduce self-reported and physiological indices of anxiety in individuals exposed to a social stressor (Heeren, Reese, McNally, & Philippot, 2012). Implications of the CogBIAS hypotheses for SPS are discussed in Section 3.

1.8 Implications for research on SPS

The various theories of environmental sensitivity and their interrelationships are described in Fig. 1. What conclusions and future directions can

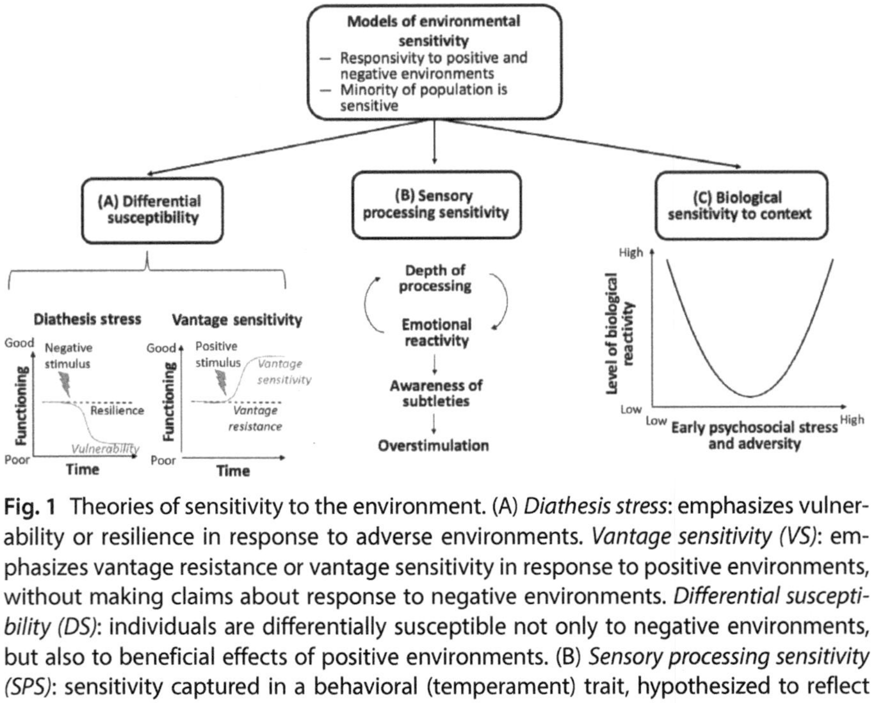

Fig. 1 Theories of sensitivity to the environment. (A) *Diathesis stress*: emphasizes vulnerability or resilience in response to adverse environments. *Vantage sensitivity (VS)*: emphasizes vantage resistance or vantage sensitivity in response to positive environments, without making claims about response to negative environments. *Differential susceptibility (DS)*: individuals are differentially susceptible not only to negative environments, but also to beneficial effects of positive environments. (B) *Sensory processing sensitivity (SPS)*: sensitivity captured in a behavioral (temperament) trait, hypothesized to reflect increased depth of processing, awareness of subtleties, emotional reactivity, and ease of overstimulation. Later theorizing suggests that depth of processing, in interaction with emotional reactivity, is the core underlying component, leading to greater awareness of subtleties and ease of overstimulation. (C) *Biological sensitivity to context (BSC)*: emphasis on sensitivity as a biological property indexed by heightened reactivity in stress response system. Hypothesizes biological reactivity to emerge from both highly stressful and highly protected early social environments. Models (A), (B), and (C) all describe individual differences in sensitivity to both negative and positive environments, and all hypothesize heightened sensitivity or reactivity to be present in a minority of the population, for evolutionary reasons. *(© Taken from Greven, C. U., Lionetti, F., Booth, C., Aron, E. N., Fox, E., Schendan, H. E., … Homberg, J. (2019). Sensory processing sensitivity in the context of environmental sensitivity: A critical review and development of research agenda. Neuroscience and Biobehavioral Reviews, 98, 287–305. http://dx.doi.org/10.1016/j.neubiorev.2019.01.009, published creative common license Attribution 4.0 International (CC BY 4.0).)*

be drawn from putting SPS in the context of these other theories? One question pertains to whether SPS is better conceptualized as reflecting developmental plasticity or immediate reactivity. SPS is proposed to be a relatively stable trait, shown to moderate adult psychopathological outcomes as a function of childhood experiences (Greven et al., 2019). In addition, it is conceivable that the extent to which someone scores higher or lower on SPS scales, is determined by childhood experiences shaping the level of the SPS trait, which then stays more stable throughout adult development. Further, in adolescents, interindividual differences in SPS are around 50% attributable to genetic factors and 50% to environmental factors (Assary, Zavos, Krapohl, & Pluess, 2019). It is also known, from research outside SPS, that traumatic experience can make someone more sensitive to sensory stimuli (Robinson & Brown, 2016). Hence, both developmental plasticity and immediate reactivity are plausible for SPS. Yet, there is a need for clearer differentiation of these two aspects.

Another issue, which is common to SPS theory, BSC, and DS theory, is the assumption that a minority of the population are sensitive, for biological reasons. The question then is, where is the appropriate cutoff to define those who are "highly sensitive." It is also a question whether we need or will find such a cutoff. Two recent studies examined this question in children and adults, using latent class analyses to uncover three sensitivity classes: 20%–35% were highly sensitive, 41%–47% moderately sensitive, and 20%–35% were low in sensitivity (Lionetti et al., 2018; Pluess et al., 2017). Whether these sensitivity classes represent qualitative differences (differing in kind, e.g., in the outcomes they correlate with) or quantitative (differing in extent, e.g., in the magnitude of the association) has not conclusively been addressed yet. Quantitative differences would suggest that the groups reflect an underlying continuum of interindividual differences in SPS traits. That is, everyone is placed somewhere on the sensitivity spectrum, rather than there being two or three qualitatively distinct groups of individuals high-, medium-, low-sensitive individuals.

For many other traits, there is strong and conclusive evidence that these traits are best seen as quantitative continua rather than qualitative taxons; in other words, reflecting continua rather than categories. For example, most personality traits are reflected on a continuum (e.g., there is no particular cutoff that makes a person an introvert or extravert) (Haslam, Holland, & Kuppens, 2012). The same is true for prevalent disorders such as attention-deficit hyperactivity disorder (ADHD), autism spectrum disorder (ASD), and even schizophrenia, which are considered quantitative traits

(Plomin, Haworth, & Davis, 2009; Zavos et al., 2014). The idea proposed, is that these traits/disorders are part of a continuum, and individuals all lie somewhere on the ADHD and ASD spectrum, ranging from low, medium, to high. The implication of quantitative variability in traits is that the high end of a trait, often referred to as "abnormal," is actually part of "normal" species variation, ultimately serving evolutionary fitness through greater diversity in strategies for survival and reproduction. Coming back to the examples from the mental health literature, ASD and ADHD have been placed in the context of the neurodiversity movement, which acknowledges the important individual contributions of the unique traits inherent to ASD and ADHD to society (Baron-Cohen, 2017). There is also increasing acknowledgment of the positive sides of these disorders, such as increased creativity and energy for ADHD, and a detail-focused cognitive style for ASD (Greven, Buitelaar, & Salum, 2018; Happe & Booth, 2008; Sedgwick, Merwood, & Asherson, 2018).

A third implication from the mentioned theories lies in the basis they provide for conducting biological research on SPS, which is discussed in detail in Chapter 5 of this book. BSC theory focuses on underlying physiological differences in reactivity of the stress-response systems, such as cortisol production, arterial pressure, immune reactivity (Boyce & Ellis, 2005). DS theory also has a strong focus on underlying biological (e.g., genetic) mechanisms (Belsky & Pluess, 2009). This stands in contrast to SPS theory, which started with a behavioral marker of sensitivity, the SPS trait assessed via questionnaires, and only later connecting it to research on neural mechanisms that underlie expression of the trait (for review, see Acevedo et al., 2018). Across several scientific disciplines relevant to mental health research, there has in recent years been a move from top-down to bottom-up approaches for understanding the etiologies of mental disorders. For example, studying ADHD in a top-down manner, one may start by assessing behavioral symptoms (e.g., by questionnaire or diagnostic interview) and then studying how genes, environments, and neurobiology contribute to these symptoms. In contrast, the bottom-up approach would start with basic dimensions of behavior linked to ADHD such as attention, reward-processing, impulsivity, and inhibition; and from there would subsequently study the relevance to (symptoms of) the disorder. This is referred to as the Research Domain Criteria (RDoC) approach, which has gained popularity in mental health research (Garvey, Avenevoli, & Anderson, 2016; Insel, 2014). The reason for developing this approach is that research on the etiologies of mental disorders that started at the level of behavior or diagnosis, has been mixed and often yielded small effect sizes, although many advances have been made. Research may have been hampered by the fact that mental health continua such as ADHD and ASD are characterized by substantial heterogeneity. That is, no person with ASD or ADHD is alike, and that in some people, ADHD or ASD, despite looking the same at the level of behavior, varies according to biological mechanisms. A famous quote is: "If you've met one individual with autism, you've met one individual with autism" (AutismSpeaks, 2018).

SPS is conceptualized as a temperament trait and not a disorder, yet SPS research can inform and be informed by these approaches in mental health research. For example, the RDoC approach is also relevant to temperament traits such as SPS. It suggests that a two-pronged approach can be helpful: one which starts with assessing SPS by questionnaires or interviews, and is characterized by the behaviors described by SPS theory, notably depth-of-processing and emotional reactivity, including sensitivity to others' moods. A second approach would start by studying biological mechanisms relevant to SPS (e.g., basic dimensions of perception, attention, emotional reactivity) and how they relate to environments. Further, there is likely substantial heterogeneity underlying high sensitivity. That is, two highly sensitive persons may express the trait differently (whether it is the degree, style, or things that they are particularly sensitive to). Also, high sensitivity may have different etiologies in different people, a topic that remains to be examined by research.

2 Empirical evidence for SPS as a marker for environmental sensitivity

Having reviewed different theories of environmental sensitivity, and having placed SPS in the context of these theories, Section 2 of this chapter will describe empirical evidence that SPS is indeed a marker of sensitivity to the environment, in line with diathesis-stress, differential susceptibility, or vantage sensitivity.

2.1 Evidence favoring diathesis-stress

Observational studies. Early work on SPS was based on Aron and Aron's (1997) identification of two groups of highly sensitive (i.e., high in SPS) individuals: The first, smaller group (about one-third of participants) had significantly higher scores on outcomes, including social introversion, negative emotionality, and having had an unhappy childhood compared to those in the second, larger group (about two-thirds of participants). The second, larger group was mostly similar on these outcomes to individuals who were not highly sensitive; with the exception of daylight sensitivity and alcohol sensitivity, which was higher for both groups (i.e., the smaller and larger group) of highly sensitive individuals compared to those less sensitive. Importantly, the two highly sensitive groups did not differ in level of sensitivity. The authors hypothesized that these differences in outcomes between the high sensitivity groups could be explained by the moderating role of high SPS with respect to links between childhood environment and developmental outcomes. This hypothesis was supported, such that when parental environment was positive, supportive, and nurturing, there were few differences between individuals who were highly sensitive and compared to those who were not. However, when parental environment was harsh or poor, highly sensitive individuals reported greater unhappiness in their childhood. This first study on SPS was therefore consistent with the view that SPS might be a vulnerability factor, reflecting diathesis-stress. Yet, the study focused on items related to negative childhood experiences, whereas items on positive childhood experiences would be needed for a full test of diathesis-stress vs. differential susceptibility.

Two additional observational studies were also more consistent with SPS reflecting diathesis-stress rather than differential susceptibility. In a sample of college students, Liss and colleagues found that higher SPS was related to greater anxiety and depression, over and above parental care (Liss, Timmel, Baxley, & Killingsworth, 2005). This suggests that higher sensitivity is, on average, related to higher levels of anxiety and depression independent of how warm, affectionate, intrusive, or controlling parents were to their child in the first 16 years of life. Further, the study found that when parental care was low, highly sensitive individuals, relative to nonhighly sensitive individuals, had significantly higher depression scores (Liss et al., 2005). However, when parental care was high, there was no significant difference between highly sensitive and nonhighly sensitive individuals, in line with diathesis-stress.

The other observational study of SPS involved a sample of adults, which showed that highly sensitive individuals, relative to nonhighly sensitive individuals, had significantly lower life satisfaction when childhood experiences were negative (Booth, Standage, & Fox, 2015). However, there were no differences in life satisfaction between highly and nonhighly sensitive individuals when childhood experiences were positive (Booth et al., 2015). The authors provided several explanations, which could explain why their results were in line with diathesis-stress rather than the expected differential susceptibility. One explanation is that most people reported positive childhood experiences, which led to a skewed distribution. Another one is retrospective, self-report of childhood experiences, which could have led to an underpowered design. The use of experimental manipulations of positive as well as negative experiences would be able to overcome biases due to retrospective report.

2.2 Evidence favoring differential susceptibility and vantage sensitivity

Observational studies. There have also been several observational studies more in line with SPS acting as a differential susceptibility factor, moderating sensitivity to the environment in a for-better and for-worse fashion. A study by Aron, Aron, and Davies (2005) found that highly sensitive adults were more likely to be shy and higher in negative affect if they had reported adverse childhood experiences. However, there was some evidence for a crossover interaction, such that highly sensitive individuals with positive childhood experiences, were no more likely, and possibly even less likely, to be shy or high in negative affect than less sensitive individuals. This provides tentative evidence that, as well as being more affected by negative childhood experiences, highly sensitive individuals may benefit more from positive childhood experiences in terms of showing lower levels of shyness and negative affect.

A study by Slagt, Dubas, van Aken, Ellis, and Dekovic (2017) examining kindergarten children followed up over one-year, found that highly sensitive children showed the biggest decrease in levels of externalizing behavior when negative parenting decreased, yet they showed the largest increase in externalizing behavior when negative parenting increased. Further, highly sensitive children showed the largest decreases in externalizing behavior when high levels of positive parenting were maintained, but showed the highest increase in externalizing behavior when positive parenting decreased. These findings support differential susceptibility. In addition, the study also revealed some support in favor of vantage sensitivity. For example, the study found that highly sensitive children showed the lowest externalizing behavior when high levels of positive parenting were maintained, but levels of externalizing behavior were similar when positive parenting decreased. However, the study found no evidence that SPS interacts with negative and positive parenting to predict prosocial behavior. A strength of this study was the use of a longitudinal design and the assessment of positive and negative parenting, and child outcomes, allowing for a better test of which model of environmental sensitivity best applies to SPS.

Experimental manipulation. Observational studies are highly valuable in studying the effects of environments when manipulation of the environment is not possible or unethical (e.g., it would not be ethical to expose children to an adverse childhood for research purposes). Yet, experimental manipulation of environments, when possible, has the advantage of allowing stronger inferences about the causal influences of positive and negative environments. A study using experimental manipulation (Lionetti et al., 2018) showed happy and sad video clips to adults, and assessed their mood before and after this mood induction. The study found that highly sensitivity individuals benefitted the most from positive mood induction, but were not significantly more reactive to sad mood induction than less sensitive individuals. Thus, this study provided support for vantage sensitivity. However, there was a nonsignificant trend such that highly sensitive individuals were also more reactive to negative mood induction, suggesting that the study may have been underpowered to detect differential susceptibility. Also, other laboratory studies have shown that highly sensitive (vs less sensitive) individuals are indeed more susceptible to negative effects of stress or harsh environments, as discussed in Chapter 4 of this book.

Behavior observation. Behavior observation, usually done by a trained rater unrelated to the participant being assessed, comes with the advantage of being more objective, compared to ratings completed by the participant themselves, for instance via questionnaires. Recently, a behavior observation tool for rating SPS in preschoolers was developed, called the Highly Sensitive Child (HSC)-Rating System (Lionetti, Aron, Aron, Klein, & Pluess, 2017). Based on this system, it has been shown that higher SPS in children is related to more behavioral problems when early parenting is negative, but may also be related to better social competence in the case of positive parenting styles. This is consistent with differential susceptibility.

Intervention studies. Notably, SPS has consistently been shown to moderate the impact of intervention and prevention programs on mental health and behavior, consistent with vantage sensitivity models. A study applying a resilience-promoting school-based depression prevention program in a sample of girls from an economically deprived background (Pluess & Boniwell, 2015) found that girls with high HSC scores, compared to those with low HSC scores, had significantly lower depression levels at 6- and 12-months follow-up, although high- and low-scoring groups did not differ in their depression levels at baseline. Furthermore, the high HSC group had significantly lower depression scores at 12-month follow-up, compared with the control group (which did not undergo the prevention program).

However, the low SPS group did not differ from the control group. Thus, although the prevention program was not found to be effective across the whole sample, it was effective in showing sustained benefits in the high SPS group (Pluess & Boniwell, 2015). This finding underlies the importance of studying interindividual differences in response to environments rather than just mean differences, made in Section 1.6.

Results favoring vantage sensitivity were also found for a school-based antibullying intervention in a large randomized controlled trial ($N = 2042$ children) from 13 schools in Italy (Nocentini, Menesini, & Pluess, 2018). The results of the study were such that boys with high SPS levels showed a reduction in levels of victimization and internalizing symptoms, but boys scoring low on SPS showed no response to the intervention. In highly sensitive boys, the effect of the intervention was more than twice as large as the average effect in the total sample. A possible explanation for this finding is that highly sensitive individuals may process and internalize information more deeply, as suggested by SPS theory and research, thus allowing them repeated application of the intervention or prevention strategies (Greven et al., 2019).

2.3 Other evidence for SPS as a marker of environmental sensitivity

Other lines of evidence linking higher SPS to greater sensitivity to one's environmental surroundings stem from research on the neurobiological and genetic basis of SPS. We refer to Chapter 5, which includes a detailed summary on these issues.

2.4 Conclusion

Diathesis-stress, differential susceptibility, vantage sensitivity, and SPS theory all have commonalities and unique perspectives to explain why some individuals show greater sensitivity and responsiveness to the environment. The emerging picture appears to be complex, and in fact there is support for all three mechanisms. It may well be that all three models are true, but that the applicable model may differ depending on the type of environmental exposure, and even for different highly sensitive individuals. This would be consistent with the different sensitivity types, described in the metaframework of environmental sensitivity (see Section 1.6), that develop depending on early environmental exposure. Such sensitivity types or subgroups remain to be examined empirically in relation to SPS. Future studies may also apply longitudinal prospective designs, experimental manipulation of environments, and use of positive as well as negative environmental exposure and outcome variables in order to allow a more thorough test of diathesis-stress, differential susceptibility, and vantage sensitivity. It would also be helpful to learn more about how SPS relates to sensitivity to the environment in everyday life, to add to the body of work with observational and laboratory-based methods. One such novel methodology is called "experience sampling" which allows the assessment of everyday microstressors and daily uplifts in a real-world setting, by using a signal on the smart phone or similar device to remind participants to provide ratings of their moods, thoughts, activities, and locations several times a day or over several days. This type of methodology would provide a rich dataset, which would allow us to learn more about how sensitivity to the environment leads to better or worse outcomes, which is vital for informing prevention of potential negative outcomes and fostering human potential and flourishing among highly sensitive individuals.

3 Implications and future directions

For a long time, research has focused on a deficit-based biomedical models of sensitivity to the environment (e.g., diathesis stress). More recently, there is a slow and increasing shift to a more balanced view that fosters strengths and investigates conditions that promote thriving, without ignoring possible risk for vulnerabilities and psychopathologies in individuals with high SPS and sensitivity to the environment. What existing scientific research shows is that SPS indexes much more than vulnerability to negative environments, but also responsivity to positive experiences as well.

What are the implications for variations in sensitivity to the environment for our society? Roughly 20%–30% of the population are considered high SPS. The implication is that, if we could intervene and foster potential in high SPS individuals, we would be able to shift the entire distribution of stress-related problems and adaptive functioning in a salutary direction, with the potential to yield large improvements at the population level. Research evidence suggests that higher SPS is linked to a range of psychopathologies, especially internalizing and stress-related problems such as higher levels of anxiety, depression, and social anxiety; lower levels of subjective happiness and life satisfaction; and higher stress-management difficulties including emotion regulation difficulties, increased levels of stress, physical symptoms resulting from ill health, greater work displeasure, and need for recovery from work-related factors (Greven et al., 2019). Yet, studies also make clear that higher SPS is linked to a range of positive outcomes, such as being more creative, empathic, and having good mental health when environments are supportive (Andresen, Goldmann, & Volodina, 2018; Bakker & Moulding, 2012; Evers, Rasche, & Schabracq, 2008; Greven et al., 2019; Jonsson, Grim, & Kjellgren, 2014; Miriam Liss, Mailloux, & Erchull, 2008; Meredith, Bailey, Strong, & Rappel, 2016; Neal, Edelmann, & Glachan, 2002; Sobocko & Zelenski, 2015). SPS may therefore be an overlooked factor in resolving the large individual, social, and economic burden of stress-related problems (Cuijpers, Beekman, & Reynolds, 2012; OECD, 2012; Samele et al., 2018; Whiteford, Ferrari, Degenhardt, Feigin, & Vos, 2015), preserving human capital, and promoting human flourishing.

Also, the evidence reviewed in this chapter, suggests that it would be particularly important and fruitful to prevent the development of potential negative outcomes and also to foster positive development among highly sensitive individuals. An important piece of background information is the existence of a negative attentional bias, also negativity bias: the evolutionarily conserved tendency of all individuals to attend to, use and learn from negative information much more than positive information (Vaish, Grossmann, & Woodward, 2008). In other words, whether sensitive or not, we all tend to be more attentive and responsive to negative (relative to positive) environmental information. In line with the thinking of the CogBIAS hypothesis (described in Section 1.7), the implication is that it may be particularly important to prevent the development of "toxic" cognitive biases in individuals with a genetic predisposition to being highly sensitive, and foster the development of more flexible attentional orientation, to selectively process positive information in supportive environments. Those with low genetic disposition to sensitivity would have a natural protective factor in adverse situations, but would need more intensive intervention to foster the development of cognitive processing styles that promote well-being. As a result, it is the environmental conditions that would influence whether high SPS is expressed as positive, strength or adverse development.

An advantage is that SPS can be observed at the level of behavior and can therefore relatively easily be identified to allow targeted prevention. However, measures of SPS need to improve beyond questionnaires, as questionnaires are subjective. In younger children, the HSC-Rating System (Lionetti et al., 2017), which uses objective behavioral observations by parents or teachers, is a good starting point (see Section 2.2). In adults, semistructured interview-based measures would be useful, which could be developed using a similar approach as developed for other personality traits (Trull et al., 1998), and which are more objective due to including the perspective of a trainer interviewer.

The work reviewed in this chapter also has implications for a diversity-aware society. It means that we may need to consider adjusting healthcare, schools, work places, parenting, and services to interindividual differences in environmental and SPS. Importantly, this can be to the benefit of less sensitive individuals as well. For example, it allows building stronger teams through diversity, and adjusting work environments, classrooms, and treatments to meet the needs encountered. Highly sensitive individuals may be expected to be more responsive to (certain forms of) psychosocial intervention, whereas less sensitive individuals may need a more intense intervention, or different strategies. Further, adaptation to level of sensitivity goes two ways, as sensitive and less sensitive partners may need to contribute to this. For this, increasing education and awareness about interindividual differences in aspects of temperament and environmental sensitivity is key.

One unresolved question is, why, if interindividual differences in environmental sensitivity are evolutionarily-preserved, is there such an upsurge now where people have started to increasingly recognize themselves as being high in the trait? Many answers are possible, but one that we speculate about here is that the development of the SPS framework with corresponding measures (based on extensive literature studies of animals and humans, and extensive qualitative interviews with highly sensitive individuals) has contributed to an increases awareness and recognition of the trait. It may be that so many people identify themselves as having the trait because the scales developed to characterize the trait were quite thorough in taking into account the opinions and visions of highly sensitive individuals. Further, SPS may be particularly popular as it is observable as a behavioral trait and "measurable" by questionnaires, as well as summarized in popular books (e.g., Aron, 2017).

4 Conclusion

To conclude, this chapter describes several theories explaining what makes some individuals more environmentally sensitive than others, places SPS in the context of these theories, and outlines how SPS not only increases sensitivity to negative, but also positive environments. We suggest that SPS moderates sensitivity to positive and negative environmental influences. As such, SPS may be relevant to resolving important societal issues, as its identification may allow prevention of potential adverse outcomes (e.g., stress-related problems), as well as being relevant to fostering the positive side of the SPS trait (e.g., creativity). Ultimately, understanding and harnessing SPS in society may prove to be important for increasing mental health, work place productivity, personal relationships, and the preservation of human capital.

References

Acevedo, B. P., Aron, E. N., Aron, A., Sangster, M. D., Collins, N., & Brown, L. (2014). The highly sensitive brain: An fMRI study of sensory processing sensitivity and response to others' emotions. *Brain and Behavior: A Cognitive Neuroscience Perspective*, *4*(4), 580–594.

Acevedo, B. P., Aron, E. N., Pospos, S., & Jessen, D. (2018). The functional highly sensitive brain: A review of the brain circuits underlying sensory processing sensitivity and seemingly related disorders. *Philosophical Transactions of the Royal Society, B: Biological Sciences*, *373*(1744), https://doi.org/10.1098/rstb.2017.0161.

Acevedo, B. P., Jagiellowicz, J., Aron, E. N., Aron, A., & Marhenke, R. (2017). Sensory processing sensitivity and childhood quality's effects on neural responses to emotional stimuli. *Clinical Neuropsychiatry*, *14*(6), 359–373.

Andresen, M., Goldmann, P., & Volodina, A. (2018). Do overwhelmed expatriates intend to leave? The effects of sensory processing sensitivity, stress, and social capital on expatriates' turnover intention. *European Management Review*, *15*(3), 315–328.

Aron, E. N. (2017). *The highly sensitive person: How to thrive when the world overwhelms you* (Thorsons Classics edition). London: Harper Thorsons.

Aron, E. N., & Aron, A. (1997). Sensory-processing sensitivity and its relation to introversion and emotionality. *Journal of Personality and Social Psychology*, *73*(2), 345–368.

Aron, E. N., Aron, A., & Davies, K. M. (2005). Adult shyness: The interaction of temperamental sensitivity and an adverse childhood environment. *Personality and Social Psychology Bulletin*, *31*(2), 181–197. https://doi.org/10.1177/0146167204271419.

Aron, E. N., Aron, A., & Jagiellowicz, J. (2012). Sensory processing sensitivity: A review in the light of the evolution of biological responsivity. *Personality and Social Psychology Review*, *16*(3), 262–282. https://doi.org/10.1177/1088868311434213.

Assary, E., Zavos, H. M., Krapohl, E., & Pluess, M. (2019). Heritability of environmental sensitivity and genetic overlap with the Big-Five personality traits. Available from: https://www.jaacap.org/article/S0890-8567(16)30705-5/fulltext (under review).

AutismSpeaks. (2018). *10 Inspiring quotes from people with autism*. Retrieved from: https://www.autismspeaks.org/life-spectrum/10-inspiring-quotes-people-autism.

Bakker, K., & Moulding, R. (2012). Sensory-processing sensitivity, dispositional mindfulness and negative psychological symptoms. *Personality and Individual Differences*, *53*(3), 341–346.

Baron-Cohen, S. (2017). Editorial perspective: Neurodiversity—A revolutionary concept for autism and psychiatry. *Journal of Child Psychology and Psychiatry*, *58*(6), 744–747. https://doi.org/10.1111/jcpp.12703.

Belsky, J. (2000). Conditional and alternative reproductive strategies: Individual differences in susceptibility to rearing experience. In J. Rodgers, D. Rowe, & W. Miller (Eds.), *Genetic influences on human fertility and sexuality: Theoretical and empirical contributions from the biological and behavioral sciences* (pp. 127–146). Boston: Kluwer.

Belsky, J., & Pluess, M. (2009). Beyond diathesis stress: Differential susceptibility to environmental influences. *Psychological Bulletin*, *135*(6), 885–908. https://doi.org/10.1037/a0017376.

Booth, C., Standage, H., & Fox, E. (2015). Sensory-processing sensitivity moderates the association between childhood experiences and adult life satisfaction. *Personality and Individual Differences, 87*, 24–29. https://doi.org/10.1016/j.paid.2015.07.020.

Boyce, W. T. (2016). Differential susceptibility of the developing brain to contextual adversity and stress. *Neuropsychopharmacology, 41*(1), 142–162. https://doi.org/10.1038/npp.2015.294.

Boyce, W. T., Chesney, M., Alkon, A., Tschann, J. M., Adams, S., Chesterman, B., … Wara, D. (1995). Psychobiologic reactivity to stress and childhood respiratory illnesses: Results of two prospective studies. *Psychosomatic Medicine, 57*(5), 411–422.

Boyce, W. T., & Ellis, B. J. (2005). Biological sensitivity to context: I. An evolutionary-developmental theory of the origins and functions of stress reactivity. *Development and Psychopathology, 17*(2), 271–301.

Cuijpers, P., Beekman, A. T., Reynolds, C. F., 3rd. (2012). Preventing depression: A global priority. *JAMA, 307*(10), 1033–1034. https://doi.org/10.1001/jama.2012.271.

Ellis, B. J., Boyce, W. T., Belsky, J., Bakermans-Kranenburg, M. J., & van IJzendoorn, M. H. (2011). Differential susceptibility to the environment: An evolutionary—Neurodevelopmental theory. *Development and Psychopathology, 23*(1), 7–28. https://doi.org/10.1017/s0954579410000611.

Evers, A., Rasche, J., & Schabracq, M. J. (2008). High sensory processing sensitivity at work. *International Journal of Stress Management, 15*(2), 189–198.

Fox, E., & Beevers, C. G. (2016). Differential sensitivity to the environment: Contribution of cognitive biases and genes to psychological wellbeing. *Molecular Psychiatry, 21*(12), 1657–1662. https://doi.org/10.1038/mp.2016.114.

Garvey, M., Avenevoli, S., & Anderson, K. (2016). The National Institute of Mental Health Research domain criteria and clinical research in child and adolescent psychiatry. *Journal of the American Academy of Child and Adolescent Psychiatry, 55*(2), 93–98. https://doi.org/10.1016/j.jaac.2015.11.002.

Gottesman, I. I., & Shields, J. (1967). A polygenic theory of schizophrenia. *Proceedings of the National Academy of Sciences of the United States of America, 58*(1), 199–205.

Greven, C. U., Buitelaar, J. K., & Salum, G. A. (2018). From positive psychology to psychopathology: The continuum of attention-deficit hyperactivity disorder. *Journal of Child Psychology and Psychiatry, 59*(3), 203–212. https://doi.org/10.1111/jcpp.12786.

Greven, C. U., Lionetti, F., Booth, C., Aron, E. N., Fox, E., Schendan, H. E., … Homberg, J. (2019). Sensory processing sensitivity in the context of environmental sensitivity: A critical review and development of research agenda. *Neuroscience and Biobehavioral Reviews, 98*, 287–305. https://doi.org/10.1016/j.neubiorev.2019.01.009.

Happe, F. G., & Booth, R. D. (2008). The power of the positive: Revisiting weak coherence in autism spectrum disorders. *The Quarterly Journal of Experimental Psychology (Hove), 61*(1), 50–63. https://doi.org/10.1080/17470210701508731.

Haslam, N., Holland, E., & Kuppens, P. (2012). Categories versus dimensions in personality and psychopathology: A quantitative review of taxometric research. *Psychological Medicine, 42*(5), 903–920. https://doi.org/10.1017/S0033291711001966.

Heeren, A., Reese, H. E., McNally, R. J., & Philippot, P. (2012). Attention training toward and away from threat in social phobia: Effects on subjective, behavioral, and physiological measures of anxiety. *Behaviour Research and Therapy, 50*(1), 30–39. https://doi.org/10.1016/j.brat.2011.10.005.

Homberg, J. R., Schubert, D., Asan, E., & Aron, E. N. (2016). Sensory processing sensitivity and serotonin gene variance: Insights into mechanisms shaping environmental sensitivity. *Neuroscience & Biobehavioral Reviews, 71*, 472–483.

Insel, T. R. (2014). The NIMH Research Domain Criteria (RDoC) Project: Precision medicine for psychiatry. *The American Journal of Psychiatry, 171*(4), 395–397. https://doi.org/10.1176/appi.ajp.2014.14020138.

Jonsson, K., Grim, K., & Kjellgren, A. (2014). Do highly sensitive persons experience more nonordinary states of consciousness during sensory isolation? *Social Behavior and Personality: An International Journal, 42*(9), 1495–1506.

Lionetti, F., Aron, A., Aron, E. N., Burns, L. G., Jagiellowicz, J., & Pluess, M. (2018). Dandelions, Tulips and Orchids: Evidence for the existence of low-sensitive, medium-sensitive, and high-sensitive individuals. *Translational Psychiatry*, *8*, 24. https://doi.org/10.1038/s41398-017-0090-6.

Lionetti, F., Aron, E. N., Aron, A., Klein, D. N., & Pluess, M. (2017). *Evaluating environmental sensitivity in preschoolers: The highly sensitive child rating system*. In *Paper presented at the 18th European conference on developmental psychology, Utrecht*.

Liss, M., Mailloux, J., & Erchull, M. J. (2008). The relationships between sensory processing sensitivity, alexithymia, autism, depression, and anxiety. *Personality and Individual Differences*, *45*(3), 255–259.

Liss, M., Timmel, L., Baxley, K., & Killingsworth, P. (2005). Senory processing sensitivity and its relation to parental bonding, anxiety, and depression. *Personality and Individual Differences*, *39*, 1429–1439.

MacLeod, C., Rutherford, E., Campbell, L., Ebsworthy, G., & Holker, L. (2002). Selective attention and emotional vulnerability: Assessing the causal basis of their association through the experimental manipulation of attentional bias. *Journal of Abnormal Psychology*, *111*(1), 107–123.

Meredith, P. J., Bailey, K. J., Strong, J., & Rappel, G. (2016). Adult attachment, sensory processing, and distress in healthy adults. *American Journal of Occupational Therapy*, *70*(1). 7001250010p7001250011–7001250010p7001250018.

Merton, R. K. (1973). *The sociology of science: Theoretical and empirical investigations*. Chicago: University of Chicago Press.

Neal, J. A., Edelmann, R. J., & Glachan, M. (2002). Behavioural inhibition and symptoms of anxiety and depression: Is there a specific relationship with social phobia? *British Journal of Clinical Psychology*, *41*(4), 361–374.

Nocentini, A., Menesini, E., & Pluess, M. (2018). The personality trait of environmental sensitivity predicts children's positive response to school-based anti-bullying intervention. *Clinical Psychological Science*, https://doi.org/10.1177/2167702618782194.

OECD. (2012). *Sick on the job? Myths and realities about mental health and work*. Retrieved from: https://www.gamian.eu/wp-content/uploads/2014/05/Miranda_OECD_sick-on-the-job_presentation_Vilnius-2013.pdf.

Plomin, R., Haworth, C. M., & Davis, O. S. (2009). Common disorders are quantitative traits. *Nature Reviews. Genetics*, *10*(12), 872–878.

Pluess, M. (2015). Individual differences in environmental sensitivity. *Child Development Perspectives*, *9*(3), 138–143. https://doi.org/10.1111/cdep.12120.

Pluess, M., Assary, E., Lionetti, F., Lester, K. J., Krapohl, E., Aron, E. N., & Aron, A. (2017). Environmental sensitivity in children: Development of the highly sensitive child scale and identification of sensitivity groups. *Developmental Psychology*, https://doi.org/10.1037/dev0000406.

Pluess, M., & Belsky, J. (2011). Prenatal programming or postnatal plasticity? *Development and Psychopathology*, *23*(1), 29–38. https://doi.org/10.1017/S0954579410000623.

Pluess, M., & Belsky, J. (2013). Vantage sensitivity: Individual differences in response to positive experiences. *Psychological Bulletin*, *139*(4), 901–916. https://doi.org/10.1037/a0030196.

Pluess, M., & Boniwell, I. (2015). Sensory processing sensitivity predicts treatment response to a school-based depression prevention program: Evidence of vantage sensitivity. *Personality and Individual Differences*, *82*, 40–45.

Robinson, C., & Brown, A. M. (2016). Considering sensory processing issues in trauma affected children: The physical environment in children's residential homes. *Scottish Journal of Residential Child Care*, *15*(1), 6–18.

Samele, C., Lees-Manning, H., Zamperoni, V., Goldie, V., Thorpe, L., Wooster, E., … Grange, R. (2018). *Stress: Are we coping?* Mental Health Foundation. *https://www.mentalhealth.org.uk/publications/stress-are-we-coping*.

Sameroff, A. J. (1983). Developmental systems: Context and evolution. In P. Mussen (Ed.), Vol. 1. *Handbook of child psychology* (pp. 237–294). New York: Wiley.

Sedgwick, J. A., Merwood, A., & Asherson, P. (2018). The positive aspects of attention deficit hyperactivity disorder: A qualitative investigation of successful adults with ADHD. *Attention Deficit and Hyperactivity Disorders*, https://doi.org/10.1007/s12402-018-0277-6.
Slagt, M., Dubas, J. S., van Aken, M. A., Ellis, B. J., & Dekovic, M. (2017). Sensory processing sensitivity as a marker of differential susceptibility to parenting. *Developmental Psychology*, https://doi.org/10.1037/dev00004 [Online first publication].
Sobocko, K., & Zelenski, J. M. (2015). Trait sensory-processing sensitivity and subjective well-being: Distinctive associations for different aspects of sensitivity. *Personality and Individual Differences*, *83*, 44–49.
Trull, T. J., Widiger, T. A., Useda, J. D., Holcomb, J., Doan, B.-T., Axelrod, S. R., … Gershuny, B. S. (1998). A structured interview for the assessment of the five-factor model of personality. *Psychological Assessment*, *10*(3), 229.
Vaish, A., Grossmann, T., & Woodward, A. (2008). Not all emotions are created equal: The negativity bias in social-emotional development. *Psychological Bulletin*, *134*(3), 383–403. https://doi.org/10.1037/0033-2909.134.3.383.
Whiteford, H. A., Ferrari, A. J., Degenhardt, L., Feigin, V., & Vos, T. (2015). The global burden of mental, neurological and substance use disorders: An analysis from the Global Burden of Disease Study 2010. *PLoS One*, *10*(2), e0116820. https://doi.org/10.1371/journal.pone.0116820.
Wolf, M., van Doorn, G. S., & Weissing, F. J. (2008). Evolutionary emergence of responsive and unresponsive personalities. *Proceedings of the National Academy of Sciences of the United States of America*, *105*(41), 15825–15830. https://doi.org/10.1073/pnas.0805473105.
Zavos, H. M., Freeman, D., Haworth, C. M., McGuire, P., Plomin, R., Cardno, A. G., & Ronald, A. (2014). Consistent etiology of severe, frequent psychotic experiences and milder, less frequent manifestations: A twin study of specific psychotic experiences in adolescence. *JAMA Psychiatry*, *71*(9), 1049–1057. https://doi.org/10.1001/jamapsychiatry.2014.994.

CHAPTER 4

Health and social outcomes in highly sensitive persons

Jadzia Jagiellowicz[a], Sheerin Zarinafsar[b], Bianca P. Acevedo[b]
[a]Department of Psychology, Stony Brook University, Stony Brook, NY, United States
[b]Department of Psychological and Brain Sciences, University of California, Santa Barbara, CA, United States

Contents

1 Theories of SPS 75
2 Health and social outcomes in children and adolescents 76
  2.1 Depression and anxiety in highly sensitive children 76
  2.2 Moderators of social adjustment 80
  2.3 Peer relationships and adolescents 86
3 Health and social outcomes in adults 88
  3.1 Physical health 89
  3.2 Mental health 89
  3.3 Life and work satisfaction 92
  3.4 Social outcomes 95
4 Conclusion 97
References 98

1 Theories of SPS

Individuals vary in their sensitivity to environmental stimuli, and particularly in response to novel environments—with some being bold, aggressive, and impulsive, others being more moderate, and a significant minority appearing to be generally less aggressive and more cautious. The difference in reactions to stimuli in the environment is linked to how strongly the individual is able to perceive and process incoming information (Pluess, 2015). Such differences in sensing and processing environmental stimuli may represent the underlying common denominator linking the various models which comprise *environmental sensitivity* (ES), of which sensory-processing sensitivity (SPS; Aron, Aron, & Jagiellowicz, 2012) is one model. Specifically, SPS, measured with the Highly Sensitive Person (HSP) Scale (Aron & Aron, 1997), proposes an underlying phenotypic temperament trait characterized by deeper cognitive processing of stimuli, emotional reactivity, ease of overstimulation, and a greater awareness of environmental subtleties, including others' affective

The Highly Sensitive Brain
https://doi.org/10.1016/B978-0-12-818251-2.00004-7

states (Acevedo et al., 2014; Aron et al., 2012). Individual differences in ES/SPS are determined by both biological factors and environmental influences, including the early-childhood developmental environment (Belsky & Pluess, 2009). This chapter focuses specifically on how SPS affects health and social outcomes in childhood and adulthood.

2 Health and social outcomes in children and adolescents

2.1 Depression and anxiety in highly sensitive children

Some of the negative mental health outcomes related to high sensitivity in childhood include anxiety, depression, and externalizing behaviors, with the bulk of the literature reporting on the development of social anxiety and depression (see reviews by Degnan & Fox, 2007; Fox, Henderson, Marshall, Nichols, & Ghera, 2005). The study of negative mental health outcomes has been conducted for decades, mostly within the field of developmental psychology. According to developmental psychologists, a number of constructs all reflect similar conceptualizations to that of SPS, characterized by an individual's response to uncertainty or novelty (Degnan & Fox, 2007). The constructs have different names for the same behavioral patterns observed as the child progresses through developmental stages. For example, a common term utilized in developmental research is "reactivity" in infants, "inhibitedness" or "behavioral inhibition" in toddlers, and "social reticence," "social withdrawal," or "shyness" in children (see review by Degnan & Fox, 2007; Fox et al., 2005; Gunnar, 1994; Kagan, 1994; Nachmias, Gunnar, Mangelsdorf, Parritz, & Buss, 1996). Although the behavioral responses to novelty can range from activity (e.g., crying and movement of arms and legs) to inactivity (clinging to a parent, reticence to socially approach peers) in toddlers and children, there is always an underlying physiological response, thus coining the term "reactivity" (Boyce & Ellis, 2005; Gannon, Banks, Shelton, & Luchetta, 1989; Kagan, Reznick, & Snidman, 1987). Thus, differently named constructs describe physiological arousal to novelty, hypothesized to be related to signs of activation of the reticular activation system and/or the sympathetic system of the nervous system. It's interesting to note that as children develop, the behaviors related to high sensitivity become more apparent in their social behaviors and psychopathology (Fox, Henderson, Rubin, Calkins, & Schmidt, 2001; Kagan, Snidman, Zentner, & Peterson, 1999; Rubin, Coplan, & Bowker, 2009). Particularly, internalizing problems such as anxiety disorders (e.g., Biederman et al., 1990; Gladstone, Parker, Mitchell, Wilhelm, & Malhi, 2005; Hirshfeld-Becker et al., 2007;

van Brakel, Muris, Bogels, & Thomassen, 2006), social anxiety and social phobia (Coplan, Wilson, Frohlick, & Zelenski, 2006; Fox et al., 2001; Henderson, Marshall, Fox, & Rubin, 2004; Hirshfeld et al., 1992; Prior, Sanson, Smart, & Oberklaid, 2000; Rubin, Burgess, & Hastings, 2002; Rubin, Hymel, & Mills, 1989) become increasingly emergent in highly sensitive children.

Much of the literature on mental disorders in children, particularly social anxiety, is concerned with the impact of behavioral inhibition, which is generally a construct pertaining to toddlers, but is also used to categorize children. The pattern of anxious behaviors, social withdrawal, negative affect, and low self-esteem—all reported in the developmental literature as characteristic of behavioral inhibition—are symptoms also used to diagnose certain anxiety and mood disorders (Fox et al., 2005). Indeed, stable behavioral inhibition in children—whether determined by means of prospective longitudinal studies, retrospectively, or through cross-sectional studies—is generally associated with anxiety disorders, primarily social anxiety and social phobia (for review, see Fox et al., 2005). Longitudinal studies consistently link high levels of behavioral inhibition, from toddlerhood through middle childhood, to increased risk for social anxiety in late childhood and early adolescence (Chronis-Tuscano et al., 2009; Schwartz, Snidman, & Kagan, 1999). Children who are observed as shy—a trait related to both SPS and behavioral inhibition (Aron & Aron, 1997)—show a higher incidence of anxiety problems later on as adolescents and adults. One study showed that persistently shy children were at increased risk for anxiety, compared to children who were never shy or whose shyness ratings had decreased over time. Forty-two percent of children rated as highly shy had anxiety problems in adolescence, compared to 11% who were never rated as shy. Undergraduates and adults with social anxiety, panic disorder, and depression retrospectively reported greater childhood inhibition (Mick & Telch, 1998; Reznick, Hegeman, Kaufman, Woods, & Jacobs, 1992). Similarly, a sample of adolescents with a current diagnosis of social phobia retrospectively reported greater social avoidance in elementary school (Hayward, Killen, Kraemer, & Taylor, 1998).

Some research also suggests that SPS contributes to the development of depression in children, but only under conditions of environmental adversity (Kaufman et al., 2004; Mitchell et al., 2011; Pluess & Belsky, 2010). The development of depression among highly sensitive children is linked to the lack of social support (Kaufman et al., 2004). For example, research showed that maltreated children with high SPS and no social support had significantly higher depression scores than maltreated children with high SPS and with social support. Kaufman et al. (2004) investigated a group of 67 children removed from their caretakers due to abuse or neglect, as well as 44 community controls. Children were genotyped for the serotonin transporter protein 5-HTTLPR, of which the short-short (s/s) allele has been linked to SPS (Licht, Mortensen, & Knudsen, 2011) and depression (Caspi et al., 2003). Maltreated children with the s/s serotonin genotype and no positive social support had depression scores that were twice as high, compared with non-maltreated children with the same genotype. However, maltreated children with the s/s genotype who had positive support from a parent, relative, other adult, or friend, had only minimal increases in depression. These findings are consistent with emerging data suggesting that the negative sequelae associated with early-life stress for high SPS individuals are not inevitable, as they may be buffered by positive environmental factors, such as social support.

Animal research also suggests that parenting and SPS may interact, even prenatally, influencing mental health outcomes. For example, a study of prairie voles (Hartman & Belsky, 2018) showed that stressed voles (rodent mammals which have been used to examine the animal models of attachment bonding) cross-fostered to high-quality parental care displayed the least anxiety-like behavior and the least corticosterone reactivity of all animals. On the other hand, stressed voles cross-fostered to low-quality care displayed the most anxiety-like behavior and the most corticosterone reactivity in response to a stress-task. Additionally, while neither prenatal-stress nor rearing-condition affected oxytocin-receptor binding, high-quality rearing predicted the highest vasopressin-1a receptor binding in the amygdala, an area critical for detecting and processing emotional stimuli. Both oxytocin and vasopressin binding are implicated in social and attachment behaviors in mammals. Vasopressin-1a receptor density in the amygdala is related to social and anxiety behaviors (Carter, Grippo, Pournajafi-Nazarloo, Ruscio, & Porges, 2008). For voles in the control group, none of the outcomes were related to the rearing condition. In sum, findings from this study indicate that plasticity may affect physiology similarly, making it both a risk and a protective factor in response to rearing environment and stress.

Research has also shown that changes in positive and negative parenting also interact with SPS in predicting externalizing behavioral problems, such as conduct and attentional problems (Slagt, Dubas, van Aken, Ellis, & Dekovic, 2017). Specifically, one study showed that kindergarten children with high SPS, followed over 6 months, exhibited increases in externalizing problems in response to parenting that became more negative over time. Similarly, high SPS children showed decreases in externalizing problems when parenting became more positive over the 6-month study period.

Collectively, studies measuring SPS and environmental influences in children suggest that negative environments (such as those with high adversity, poor parenting, or lack of social support), affect highly sensitive children more strongly, often resulting in issues such as depression, anxiety, social anxiety, and externalizing behaviors through adulthood. However, the quality and availability of social support are important for promoting resiliency in maltreated children, even in the presence of a genotype expected to confer vulnerability to psychiatric disorders, such as depression. School interventions aimed at increasing resilience have also been successful with highly sensitive students, as compared to less-sensitive students (Nocentini, Menesini, & Pluess, 2018), and will be further discussed in the section on Peer Relationships later in the chapter.

2.1.1 Continuity and discontinuity in social anxiety

SPS (and related constructs) in infancy, childhood, and adolescence are thought to be based on an individual's response to uncertainty or novelty. Research on children's longitudinal profiles of inhibited behavior has focused on the heterotypic continuity between infant reactivity, toddler behavioral inhibition, childhood social withdrawal, and anxiety disorders through adulthood (Fox et al., 2005; Rothbart & Bates, 2006). Although behavioral inhibition is one of the most stable temperament traits (Fox et al., 2005; Kagan & Moss, 1962), there can be discontinuity in the trait over time, with some individuals becoming less inhibited as they reach school age and adolescence (Degnan & Fox, 2007; Schwartz et al., 1999).

A prospective, longitudinal study using a large community sample investigated the association of shy inhibited temperament with the development of anxiety problems from infancy to early adolescence (Prior et al., 2000). Since shyness has been linked to SPS (Aron, Aron, & Davies, 2005), the study results are relevant to continuity and discontinuity of anxiety in highly sensitive individuals. The study used both prospective and retrospective methods to analyze mothers' ratings of temperamentally shy children over a 13-year period in order to examine the extent to which early temperamental shyness predicted anxiety problems in later childhood and in 14-year-old adolescents. The relationship between shyness in early childhood with inhibition and anxiety became clinically meaningful starting in middle childhood (i.e., 9–10 years of age), and was predictive of anxiety problems in early adolescence when shyness persisted over time. Specifically, 42% of children rated as shy on six or more occasions over 8 surveys had anxiety problems in early adolescence, compared to 11% who were never shy. Also, 20% of anxious adolescents had been persistently shy. Thus, persistent shyness throughout childhood predicted anxiety in adolescence. Interestingly, although persistent shyness appeared to be linked to adolescent anxiety, shyness itself was not a necessary precursor of anxiety. Almost half of anxious adolescents had never been rated as shy, possibly reflecting the variety of familial and other factors, which moderate the effects of temperamental vulnerability in the development of anxiety disorders. Although this research used a proxy marker, and there is currently no longitudinal study on the stability of SPS, investigating the trajectory of SPS from childhood to adulthood would be an interesting direction for future research. Such a longitudinal study would be facilitated by the availability of two recently developed measures, i.e., the Highly Sensitive Child Scale (HSC; Weyn et al., 2019) and the Highly Sensitive Child-Rating System (HSC-RS; Lionetti, Aron, Aron, Klein, & Pluess, 2019), both described in detail in Chapter 2 of this volume.

2.2 Moderators of social adjustment

Among the factors that moderate the effects of temperament are cognitive processes, including attention, inhibitory control, and self-regulation. Parenting behaviors also play a major role in outcomes related to self-regulation, interacting with temperament, and will be discussed in Section 2.2.1. Self-regulation refers to the behaviors that an individual uses to control their behavioral and emotional reactions to sources of both positive and negative stimulation (Posner & Rothbart, 2000). Even young infants can regulate their degree of distress by directing attention away from fear-inducing objects or by sucking their fingers to distract themselves (Derryberry & Rothbart, 1997). Active self-regulation, which involves voluntary attentional control and the ability to inhibit ongoing behavior, appears later in development starting toward the end of the second year of life (Derryberry & Rothbart, 1997; Eisenberg et al., 2002). Active self-regulation can be conceptualized as effortful control and inhibitory control. Effortful control is a temperamental construct reflecting the ability to voluntarily shift attention. Inhibitory control is the ability to inhibit or override one's dominant responses in favor of subdominant responses (Eisenberg, Downs, Golberstein, & Zivin, 2009; Rothbart, Ellis, Rueda, & Posner, 2003). Relevant to SPS, one study showed that behaviorally inhibited toddlers who shifted their attention to a neutral sorting task were at lower risk for developing anxiety problems as preschoolers (White, McDermott, Degnan, Henderson, & Fox, 2011). Specifically, high levels of behavioral inhibition in toddlerhood predicted high levels of anxiety symptoms during the preschool years, but only for children exhibiting low levels of attention-shifting. This association was not detected in children showing high levels of attention-shifting. Thus, the ability to flexibly shift attention appears to protect a behaviorally inhibited child from the development of negative emotional outcomes over time. In contrast, high levels of inhibitory control were associated with increased risk for anxiety symptoms in behaviorally inhibited toddlers. Although it appears counterintuitive that inhibitory control should increase the risk for anxiety symptoms, there may be a common mechanism linking the two constructs. For example, there may be increased attention to threat, as well as rumination, both resulting from greater monitoring of performance (McDermott & Fox, 2010; Nolen-Hoeksema, 2000). Another possibility is that the increased attention being paid to negative cognitions may be related to increased anxiety symptoms (White et al., 2011). Finally, there is some evidence that exerting effortful actions to escape a stressor is associated with reductions in anxiety for some individuals (Boeke, Moscarello, LeDoux, Phelps, & Hartley, 2017; Hartley, Gorun, Reddan, Ramirez, & Phelps, 2014).

2.2.1 Parenting

Without the specific and active support of a nurturing parent, no child would thrive or even survive. Given this profound dependence on the caregiver, it is not surprising that children have a remarkable ability to respond to many different aspects of both negative and positive parenting. A large body of literature reports that highly sensitive children respond strongly to the same environmental conditions to which less sensitive children show little or no response. Specifically, early research on reactivity in infants, behavioral inhibition in toddlers, and social reticence in children was seminal in supplying evidence for the idea that childhood environment differentially affects highly sensitive individuals and influences their future development (Gunnar, 1994; Kagan, 1994; Nachmias et al., 1996).

One area that has been broadly investigated with respect to childhood environment and SPS is parenting styles. Interactions between parenting styles and child characteristics can especially influence the behavioral trajectory of highly sensitive children for the better, by training children in adaptive self-regulatory strategies; but also for the worse, by undermining children's sense of autonomy and self-confidence (Degnan & Fox, 2007;

Hastings, Zahn-Waxler, & Usher, 2007; Staub, 1979). Specifically, in one study, caregivers who were sensitive to the infant's need to disengage attention helped reduce the child's arousal before it became overwhelming (Gottman, Katz, & Hooven, 1997). This process of attention engagement and disengagement is one method by which infants learn to regulate their emotions and behavior (Fox et al., 2005). Also, by helping the child to learn self-regulatory strategies, maternal behavior can influence the child's long-term trajectory for inhibited behavior and anxiety (Degnan & Fox, 2007). Research suggests that parents who used supportive strategies may have protected their child from developing more extreme patterns of social anxiety (Wood, McLeod, Sigman, Hwang, & Chu, 2003), for example by reorienting the child's attention away from threat, and also by promoting exploration and social activity (Fox et al., 2005). Also, positive parenting styles/behaviors—such as reasoning with the child, providing warmth, support, and autonomy—promote empathy and prosocial behavior among children (Eisenberg, Damon, & Lerner, 2006, review). In contrast, intrusive and harsh parenting result in noncompliant and nonexpressive tendencies in children (Whiteside-Mansell, Bradley, Tresch Owen, Randolph, & Cauce, 2003).

More recently, genetic work has also provided support for the interactions between parenting and children's health outcomes. Numerous studies report interactions between parenting and the short-allele of the 5-HTTLPR gene, which has been associated with SPS (Fox et al., 2005; Hankin et al., 2011; Kochanska, Kim, Barry, & Philibert, 2011; Kumsta et al., 2010). The short allele of the 5-HTTLPR gene has been shown to moderate the effects of maternal responsiveness on children's moral internalization (Kochanska et al., 2011) and the influence of supportive parenting on a child's positive affect (Hankin et al., 2011). In both examples, those carrying the 5-HTTLPR short allele had the least adaptive outcomes under less favorable conditions and the most adaptive outcomes under more favorable conditions (when compared with children not carrying the polymorphism). Children with the short/short allele who were tested at 14 months and whose mothers reported low social support were more inhibited with peers at 84 months of age (Fox et al., 2005), suggesting a negative outcome bias under less favorable parenting conditions for children with the short/short 5-HTTLPR variant.

Children with SPS (i.e., more negative emotionality in infancy, toddlerhood, and childhood) are more adversely affected by unresponsive parenting and benefit substantially more from responsive parenting, in comparison to children with low SPS (Belsky, Hsieh, & Crnic, 1998; Klein Velderman,

Bakermans-Kranenburg, Juffer, & van IJzendoorn, 2006; Morrell & Murray, 2003; Obradovic, Bush, Stamperdahl, Adler, & Boyce, 2010; Belsky & Pluess, 2009). For instance, with respect to negative emotionality, Morrell and Murray (2003) showed that it was only highly distressed and irritable 4-month-old boys with coercive and rejecting mothering who continued to show evidence, 5 months later, of emotional and behavioral dysregulation—specifically, conduct disorder. Similarly, Belsky et al. (1998) observed that infants who scored high in negative emotionality at 12 months of age and who experienced the least supportive parenting across their second and third years of life, scored highest on externalizing problems. Feldman, Greenbaum, and Yirmiya (1999) found that 9-month-olds scoring high on negativity, and who showed low levels of synchrony in mother-infant interactions, manifested more noncompliance during clean-up at age two. However, when the same sensitive infants showed mutually synchronous mother-infant interactions they displayed greater self-control, compared with children that showed much less negativity as infants. Kochanska, Akson, and Joy (2007) observed that highly fearful 15-month-olds experiencing high levels of power-assertive paternal discipline were most likely to cheat in a game at 38 months; yet when cared for in a supportive manner, negatively emotional, fearful toddlers showed the most rule-compatible conduct. Klein Velderman et al. (2006) found that experimentally induced changes in maternal sensitivity exerted greater impact on the attachment security of highly negatively reactive infants than it did on other infants. In both experiments, positive environmental influences, such as enhanced maternal sensitivity, had a positive impact on "vulnerable" children.

In addition, specific types of parental behaviors are associated with child behavioral inhibition and may also be involved in the pathway to anxiety. For example, research has shown that oversolicitous, intrusive, and controlling parenting is associated with greater toddler inhibition and preschool social reticence (Rubin et al., 2002; Rubin, Cheah, & Fox, 2001; Rubin, Hastings, Stewart, Henderson, & Chen, 1997). Harsh discipline and over-intrusive behavior by parents also increased the likelihood of adolescent anxiety (Shanahan, Copeland, Costello, & Angold, 2008; van Brakel et al., 2006). For example, one study showed that children with an anxiety disorder, compared to those without an anxiety disorder, tended to have mothers that were more intrusive and overinvolved (Hudson & Rapee, 2002). In addition, greater maternal intrusive behavior was observed among children with an anxiety disorder displaying negative affect, compared to similar displays in nonaffected controls (Hudson, Comer, & Rapee, 2008).

Furthermore, parents who were oversolicitous (i.e., warm and controlling) tended to have children who maintained their behaviorally inhibited tendencies across childhood (Degnan, Henderson, Fox, & Rubin, 2008; Rubin et al., 2002). In contrast, maternal acceptance, warmth, sensitivity, and responsiveness were associated with less inhibited, more socially adaptive child behavior (Hane, Cheah, Rubin, & Fox, 2008; Park, Belsky, Putnam, & Crnic, 1997; Wood et al., 2003).

SPS was also shown to moderate the effect of childhood parenting, as measured retrospectively, on adult depressive symptoms (Liss, Timmel, Baxley, & Killingsworth, 2005). Adults scoring high on SPS reported the highest depression scores when childhood parental care had been low, while depression scores were unrelated to SPS when childhood parental quality had been high (Liss et al., 2005). Children scoring high on SPS were most responsive to changes in parenting behavior such that their externalizing behavioral problems increased when parenting became more negative, but it decreased when parenting improved (Liss et al., 2005). In some cases, a gene by environment interaction, consistent with vantage sensitivity models, was shown in response to exclusively positive experiences as a function of individual endogenous SPS factors (Pluess & Belsky, 2013). More specifically, reviews by Belsky (1997, 2005) observed that children high in negative emotionality, particularly in the early years of life, appeared to benefit disproportionately from supportive rearing environments (Boyce & Ellis, 2005).

Parental SPS can also be an important factor influencing parenting behavior, which ultimately affects the child. For example, higher levels of insecure (anxious and avoidant) parental attachment patterns were related to greater sensory sensitivity (Branjerdporn, Meredith, Strong, & Green, 2019). Highly sensitive parents are more likely to become overwhelmed by the intense sensory stimuli associated with children, such as loud or unexpected noise, interpersonal contact, mess, and tantrums. Also, high sensitivity in parents has been associated with nonoptimal parenting styles, such as permissive and authoritarian parenting behaviors. Avoidantly attached, high SPS parents displayed more permissive parenting styles; while anxiously attached, high SPS parents were more likely to respond to their children with less warmth and either controlling or permissive parenting styles. Thus, awareness of a parental SPS levels, in addition to attachment style, may assist in developing effective strategies to support the highly sensitive parent-child relationship.

Consequences of poor parenting can be seen in heightened physiological arousal in highly sensitive infants in response to stressful social and physical environments (Blunt Bugental, Martorell, & Barazza, 2003; Nachmias et al., 1996). Nachmias et al. (1996) examined cortisol levels (a measure of stress reactivity) in inhibited (i.e., highly sensitive) and uninhibited 18-month-olds, assessed for secure or insecure attachment style, in response to a challenging coping episode. Inhibited infants exhibited greater increases in cortisol than uninhibited infants, whether securely or insecurely attached. However, the cortisol response was greater in highly inhibited, insecurely attached infants compared with children who were highly inhibited but securely attached. Also, intrusive mothering behavior (i.e., those who insisted that their inhibited infants approach a novel stimulus) was associated with the child's insecure attachment and higher post-session cortisol levels.

Research investigating adverse environmental interactions with SPS has demonstrated that the resulting negative health outcomes reflect only the dark side of SPS. For example, one study showed that although children with greater psychobiological reactivity to stress (suggestive of high SPS) evidenced symptoms of poor physical health, the association was moderated by the child's environment (Boyce et al., 1995). Across two studies, researchers measured health outcomes in both highly reactive and nonreactive children: one study was conducted in an adverse home environment and the second was conducted in a stressful childcare environment (Boyce et al., 1995). In both the home and childcare environments, children with greater psychobiological reactivity to stress (vs. nonreactive children) had more respiratory illnesses, measured by immune and cardiovascular reactivity, respectively. However, children in low stress environments had lower rates of respiratory illness than less reactive peers. The authors concluded that "one plausible explanation for such a pattern of findings is that reactive children are more sensitive or more susceptible to the characteristics of the social environment" (p. 419). Less reactive children experienced no escalation in illness under stressful conditions, suggesting that only psychobiologically reactive children were susceptible to the health-altering effects of stressors and adversity. A study with adolescents found similar results (Gannon et al., 1989). Adolescents with either greater physiological arousal and/or slower recovery response to a laboratory stressor showed stronger symptoms under situations of environmental stress, compared to adolescents who were less reactive or faster to recover.

Research has also shown the positive side of SPS with respect to health outcomes in children in supportive environments. For example, Boyce et al. (1995) conducted two studies measuring incidence of respiratory disease symptoms as the primary outcome. The two studies utilized separate, nonoverlapping measures of environmental stress; one set of measures addressed stressors in childcare and the other assessed stressors at home. Study 1 used laboratory-based evaluations of cardiovascular reactivity to developmentally challenging tasks, and Study 2 measured immune reactivity to the stress of starting school. Respiratory illness incidence, the primary outcome, was measured prospectively with weekly onsite respiratory examinations in Study 1, and with biweekly parental reports of children's respiratory symptoms in Study 2. In both studies, highly reactive children exposed to low-stress environments had a lower incidence of respiratory symptoms than any other study group, despite the different measures of physiological reactivity, stress, and respiratory illness. Specifically, highly reactive children exposed to low stress environments had fewer respiratory illnesses than highly reactive children in high-stress environments and low-reactive children in both low- and high-stress environments (Boyce et al., 1995). Although environmental stress was not independently associated with respiratory illnesses in either study, the incidence of illness was related to an interaction between childcare stress and mean arterial pressure reactivity in Study 1, and to an interaction between stressful life events and immune reactivity in Study 2. In both studies, reactive children sustained higher illness rates under high-stress conditions, but lower rates in low-stress conditions, compared with less reactive peers. Consistent with theoretical models of sensitivity to the environment, which posit that ES is a plasticity factor allowing individuals to be influenced by their environment both "for better and for worse" (Belsky, 1997), such results reinforce the differential impact of a positive environment on highly reactive children, compared to those who are less physiologically reactive.

2.3 Peer relationships and adolescents

As previously discussed, highly sensitive children are predisposed to developing social anxiety and social avoidance (Coplan et al., 2006; Fox et al., 2001; Henderson et al., 2004; Prior et al., 2000; Rubin et al., 1989, 2002). Social withdrawal in childhood is related to the experience of having poor relationships with peers (Rubin, Perse, & Powell, 1985). Related to SPS, research has shown that behaviorally inhibited children may interact with others in relatively ineffective ways. For example, behaviorally inhibited children tend to use nonassertive strategies during peer interactions, and thus they are also more likely (than are their peers) to have their requests refused (Rubin & Borwick, 1984). Over time, the experience of social failure may negatively affect children's perceptions of social situations and their attributions regarding their social abilities. Thus, children may begin to interpret ambiguous social situations as threatening, and also believe that they are the cause of poor social outcomes (Goetz & Dweck, 1980; Rubin, Bukowski, & Parker, 1998). These issues may have cumulative implications over time.

Genetic evidence also shows that SPS outcomes differ as a function of early childhood environments. For example, a study investigating gene by environment interactions related to the 5-HTTLPR gene found that adolescents with a history of early institutional care and who were carrying the 5-HTTLPR short-allele, were more likely to respond with emotional problems to recent stressful life events than were adolescents homozygous for the long allele (Kumsta et al., 2010). Using a prospective-longitudinal study of 125 individuals assessed at ages 11 and 15, the study tested whether the effect of severe early institutional deprivation on adolescent emotional problems was moderated by 5-HTTLPR genotype and stressful life events in adolescence. In the severe deprivation group, carriers of the short allele (both s/s and s/l) showed the highest emotional problems, while long-allele homozygotes showed the lowest overall problem levels. Furthermore, homozygosity for the 5-HTTLPR short allele, was associated with a "for-better-and-for-worse" response in correlation with high SPS (Licht et al., 2011). Depending on whether they had experienced a high or a low number of stressful life events between 11 and 15 years of age, short-allele homozygotes had either the largest increase or largest decrease in emotional problems.

Hastings, Klimes-Dougan, Kendziora, Brand, and Zahn-Waxler (2014) measured adolescents' respiratory sinus arrhythmia (RSA), one widely studied psychophysiological marker of emotion regulation (Hastings et al., 2014), as they were shown clips that provoked sadness or fear. The study also measured maternal supportive and punishing emotion socialization. Adolescents with more arrhythmia suppression (indicating mobilizing resources to support the fight-or-flight response Porges & Furman, 2011) after watching the sad clips showed an increased risk of internalizing their emotions, especially for those with low maternal supportive emotion socialization.

As adolescents begin to spend more time with their peers, they feel a greater need to be accepted by them. For example, one research study showed greater striatal activation, a region associated with reward processing, in sensitive adolescents who thought they were going to be evaluated by a peer (Schriber & Guyer, 2015). The inhibited, sensitive adolescents also displayed a greater ability to regulate their responses in the presence of a peer. Although a potentially stressful or negative situation (i.e., being evaluated) may result in an adolescent putting too much emphasis on and adjusting their behavior to the social environment, sensitivity may also result in positive effects in supportive environments. For example, sensitive adolescents may be better able to read subtle signals from their peers, allowing them to process social situations more deeply but also making them more sensitive to negative interactions and social evaluation from peers (Somerville, 2013).

Capitalizing on the vantage sensitivity aspects of SPS, researchers have also shown that adolescents appear to benefit more from cognitive/behavioral interventions. For example, Pluess and Boniwell (2015) investigated Vantage Sensitivity as a function of SPS in adolescent girls using a universal school-based resiliency program aimed at preventing depression (Boniwell & Ryan, 2009). Girls from a disadvantaged area who were high (versus low) in SPS responded more favorably to the intervention, which was based on cognitive-behavioral therapy and positive psychology techniques (Pluess & Boniwell, 2015). Specifically, high-SPS girls showed significant decreases in depression up to 6 and 12 months after the intervention, whereas low-SPS girls showed no significant changes in depression.

In addition, a large randomized control-trial testing the efficacy of a school-based antibullying intervention among students (grades 4 and 6 students) in Italy (Nocentini et al., 2018) showed that the intervention significantly reduced bullying behaviors and improved mental health outcomes across the whole sample. However, highly sensitive boys seemed particularly responsive to the intervention, showing decreases in victimization and internalizing symptoms, but not bullying or externalizing symptoms. Highly sensitive boys did not, however, exhibit reduced bullying or externalizing symptoms. Likewise, girls with moderate sensitivity exhibited fewer victimization symptoms as a result of the intervention. Conversely, there was no significant effect for children scoring low on SPS. The authors hypothesize that highly sensitive children are more likely to be (or perceive being) victimized, and therefore were also more positively affected by a treatment-induced reduction of bullying. They explain lack of change in externalizing symptoms by the observation that sensitive children are generally less likely to manifest externalizing behaviors. These studies suggest that individual differences in SPS should be considered an important moderator of treatment response to intervention programs, and possibly other types of interventions as well.

3 Health and social outcomes in adults

What are the health and social outcomes of being highly sensitive in adulthood? It depends on a variety of factors. Generally, research suggests that in adults high sensitivity may be associated with physical and mental health issues such as pain, anxiety, and depression (Benham, 2006; Boyce et al., 1995; Engel-Yeger & Dunn, 2011; Liss et al., 2005; Jonsson, Grim, & Kjellgren, 2014; Liss, Mailloux, & Erchull, 2008; Neal, Edelmann, & Glachan, 2002). Also, SPS seems to be associated with lower levels of life and work satisfaction in adults (Booth, Standage, & Fox, 2015; Sobocko & Zelenski, 2015). In the following sections, we review the current research relating SPS with physical and mental health, and with work and life satisfaction.

3.1 Physical health

Poor physical health is reported more often in adults with high (vs. low) SPS (Ahadi & Basharpoor, 2010; Benham, 2006; Engel-Yeger & Dunn, 2011; Liss et al., 2005). Specifically, ease of excitation (EOE) and low sensory threshold (LST) aspects of SPS were shown to correlate with self-reports of poor physical health (Ahadi & Basharpoor, 2010). Similar associations between SPS and health were found, even after controlling for perceived stress and gender, in a sample of 383 American undergraduates of primarily Hispanic ethnicity (Benham, 2006). Indeed, SPS accounted for an additional 8% of the variance in health scores, after controlling for gender and stress. Controlling for gender is important in studies of ill-health, since women self-report more symptoms of poor health than do men (Benham, 2006; Engel-Yeger & Dunn, 2011). Benham (2006) postulates that more reports of ill health among highly sensitive individuals may be due to high SPS individuals being more aware of their physical/internal sensations. Also self-reports of pain may be linked to magnified sensitivity (DeGangi, 2000; Engel-Yeger & Dunn, 2011; Liss et al., 2005), and peripheral sensitivities (which will be discussed in Chapter 5). Moreover, the relationship between poor health and high SPS may result from comorbidity with psychopathology, specifically anxiety (Neal et al., 2002). For example, some studies found that sensory defensiveness was associated with anxiety (Kinnealey & Fuiek, 1999; Pfeiffer & Kinnealey, 2003). More information on clinical issues related to SPS is described in Chapters 6 and 7.

3.2 Mental health

Although previous research has focused on investigating the negative implications of SPS on mental health (i.e., increased neuroticism, anxiety and depression), gradually the field is recognizing that the neural plasticity inherent in SPS can also engender positive attributes such as enhanced creativity, empathy, awareness, and openness. Herein, we present a view that has been biased by the dominance of research on disorders, along with the few studies on the strengths of SPS with respect to cognitive functioning. First, however, we present evidence for the association between SPS and mental health issues/disorders.

3.2.1 Neuroticism

Originally, when research on SPS was in its infancy much of the psychological community did not make a distinction between SPS and Neuroticism (Aron & Aron, 1997). Neuroticism is a personality trait of the Big Five Personality inventory (Costa & McCrae, 1992) characterized by a tendency toward anxiety, negative affectivity, depression, and impulsiveness. Neuroticism has been associated with poor mental health, including anxiety, depression, and somatization (Creed, Muller, & Machin, 2001; Furnham & Cheng, 1999; Lu, 1994; Lu & Shih, 1997). Being cognizant of the relationship between Neuroticism and negative mental health outcomes, Aron and Aron (1997) conducted their early analyses of SPS data controlling for Neuroticism (Aron & Aron, 1997) to disprove the prevailing view that SPS was simply another term for negative affectivity. A second finding with regards to Neuroticism and SPS was that only individuals high in SPS who also experienced troubled childhoods reported negative emotionality, including high Neuroticism, as adults (Aron et al., 2005; Aron & Aron, 1997; Lionetti et al., 2019).

3.2.2 Anxiety and depression

As noted previously, in adults, SPS is linked to higher levels of psychopathology-related traits, including anxiety (Jonsson et al., 2014; Liss et al., 2008; Meredith, Bailey, Strong, & Rappel, 2016; Neal et al., 2002), depression (Bakker & Moulding, 2012; Liss et al., 2005; Liss et al., 2008), internalizing problems (Boterberg & Warreyn, 2016), poor stress management, difficulties in emotion regulation (Brindle, Moulding, Bakker, & Nedeljkovic, 2015), and panic disorder (Bakker & Moulding, 2012; Liss et al., 2005, 2008; Neal et al., 2002). A recent theory purports that the association of SPS with depression, anxiety, and somatic symptoms is largely driven by maladaptive thoughts and related negative emotions in response to sensory information (Wyller, Wyller, Crane, & Gjelsvik, 2018). Indeed, levels of cognitive reactivity have been suggested as a means for distinguishing between healthy and unhealthy individuals with SPS (Wyller et al., 2018).

In adults, high SPS is also associated with social anxiety disorder [defined as a marked and persistent fear of social or performance situations in which the person is exposed to unfamiliar people or to possible scrutiny by others (American Psychiatric Association, 2013)] (Hofmann & Bitran, 2006; Mick & Telch, 1998; Reznick et al., 1992). For example, adults with social anxiety, panic disorder, and depression retrospectively reported greater childhood sensitivity (Mick & Telch, 1998; Reznick et al., 1992). Similarly, research with 125 community college students showed that highly sensitive students reported more social anxiety (Hanes, 2016). Some suggest that these findings are consistent with the tendency of individuals with high SPS to be easily overwhelmed in social situations, inhibited, and to have had a history of negative interpersonal experiences (Hanes, 2016).

Researchers have also investigated what specific aspects of SPS correlate with different disorders. One study showed that EOE and LST were associated with anxiety, depression, autism (i.e., poor social skills and communication), and alexithymia symptoms (Liss et al., 2008). In contrast, aesthetic sensitivity (AES) was associated with greater attention to details and better communication. These findings are consistent with other work showing that high SPS was associated with enhanced neural responsivity to positive emotional stimuli in regions related to reward and empathy (Acevedo et al., 2014; Acevedo, Jagiellowicz, Aron, Marhenke, & Aron, 2017; Acevedo, Aron, Pospos, & Jessen, 2018). Interestingly, as aforementioned, Liss et al. (2008) reported that high AES was related to better communication, suggesting that investigation of this aspect of SPS could shed light on important biological differences and behavioral interventions.

3.2.3 Personality disorders

In addition to mood, at least one study has showed that SPS is associated with avoidant personality disorder (APD) and borderline personality disorder (BPD; Meyer, Ajchenbrenner, & Bowles, 2005). The research suggested that complex interactions among SPS, formative developmental experiences, and biases in cognitive-affective-motivational responses, give rise to the symptom patterns characteristic of these disorders. The authors administered the Structured Clinical Interview for DSM-IV Axis II screening questionnaire (SCID-II-SQ; First, Gibbon, Spitzer, Williams, & Benjamin, 1997) to measure APD and BPD features, as well as measures of temperament (including the HSP Scale measure for SPS), attachment, childhood memories, mood, and participants' reactions in various social situations that might signal the potential for rejection. They also conducted a principal components analysis, which revealed four factors in association with SPS: (1) General Sensitivity/Overstimulation (eight items, sample item: "Do you find it unpleasant to have a lot going on at once?"); (2) Adverse Reactions to Strong Sensations (four items, sample item: "Are you made uncomfortable by loud noises?"; (3) Psychological Fine-Discrimination (four items, sample item: "Are you deeply moved by arts or music?"); and (4) Controlled Harm-Avoidance (three items, including, "Do you make it a high priority to arrange your life to avoid upsetting or overwhelming situations?"). Both APD and BPD were associated with HSP scores, showing commonalities and differing pattern of results. For example, both APD and BPD correlated equally strongly with General Sensitivity/Overstimulation and Adverse Reaction to Strong Sensations. However, BPD correlated uniquely with Psychological Fine Discrimination, whereas APD correlated with Controlled Harm Avoidance. In sum, individuals with intense BPD features tended to experience stronger negative moods; more problematic attachments; and higher levels of sensitivity to particularly subtle, emotionally evocative aversive stimuli. In contrast, those with APD features were also highly sensitive but tended to interpret ambiguous social situations with a negative, rejection-implying bias, which was associated with strong negative expectancies, anxious affective responses, and avoidant tendencies.

3.2.4 Creativity

On the positive side of mental health, some research suggests that SPS is associated with creativity. For example, a very early study showed that creative individuals rated higher on measures of sensory stimulation, such as pain ratings (Martindale, 1977). Creativity also seems to be associated with higher emotionality as more creative individuals report higher negative affect and show greater sensitivity to internal affective cues (Akinola & Mendes, 2008). A recent study with a diverse sample of adults, ages 18–67, found that sensitivity (measured with the HSP scale) was positively associated with creativity (Bridges & Schendan, 2019), and that Openness (measured with the Big 5 Personality scale) was the only Big-Five factor that emerged as significantly associated with creativity. A recent review also proposes that SPS is linked to creativity through the neurosensitivity mechanisms of disinhibition and automatic orienting, as well as right hemisphere processes for global processing and integration in the default mode network (Bridges & Schendan, 2019).

3.3 Life and work satisfaction

3.3.1 Life satisfaction and happiness

Some evidence suggests that, in adults, SPS is correlated with lower levels of life satisfaction (Booth et al., 2015) and subjective happiness (Sobocko & Zelenski, 2015). Consistent with vantage sensitivity and environmental sensitivity models, the link between life satisfaction/happiness and SPS is moderated by childhood quality, environmental conditions, and the specific facet of SPS. One study found that high SPS individuals who reported having negative childhood experiences also reported lower current life satisfaction (Booth et al., 2015). However, highly sensitivity individuals who reported having had a positive childhood, had moderate levels of life satisfaction, similar to those with low SPS. Lower life satisfaction among highly sensitive individuals is also related to current life stress (Bakker & Moulding, 2012; Benham, 2006), stress-management problems, and difficulties regulating emotions (Brindle et al., 2015). In a study of almost 400 Americans, SPS was significantly correlated with perceived stress, with women reporting higher stress scores than men (Benham, 2006). In a related study, when responding to stress-related questions (measuring difficulty relaxing, nervous arousal, being easily upset, a tendency to over-react, irritability, and impatience), high SPS persons were significantly more likely to report higher stress (Bakker & Moulding, 2012).

Stress outcomes in HSPs are also related to subtle perceptions of environmental conditions, such as greater disorganization and crowding in the home. For example, Wachs (2012) asked 94 mothers to complete the HSP Scale (Aron & Aron, 1997), and a scale of home chaos measuring perceived lack of routine, disorganization, noise, and crowding in daily home life (CHAOS Scale, Matheny, Wachs, Ludwig, & Phillips, 1995). Observers also measured the level of noise, crowding, home traffic, and regularity of scheduling in the home environment. Consistent with SPS and ES models, associations between higher home traffic patterns and home disorganization with maternal perceptions of chaos in the home were significant, but only for highly sensitive mothers.

Research suggests that lower life satisfaction among highly sensitive individuals mainly results from high EOE and LST (Evans and Rothbart, 2008; Smolewska, McCabe, & Woody, 2006). Specifically, studies found that EOE and LST were significantly associated with negative affect and lower life satisfaction in adults (Booth et al., 2015; Sobocko & Zelenski, 2015). However, AES and orienting sensitivity (OS)/openness were correlated with positive affect, conscientiousness, and openness to experience (Sobocko & Zelenski, 2015).

3.3.2 *Work satisfaction among highly sensitive adults*

As with life satisfaction, high SPS is associated with a tendency toward greater work displeasure and more need for recovery time from work (Andresen, Goldmann, & Volodina, 2018; Evers, Rasche, & Schabracq, 2008).

Some evidence suggests that low work satisfaction among highly sensitive individuals is often a result of difficulty with emotion/stress regulation (Andresen et al., 2018; Evers et al., 2008). For example, according to SPS theory and the Job Demands Resources Model, the researchers predicted that negative aspects of work, (i.e., workload and emotional demands) would be associated with greater emotional exhaustion for individuals with high SPS; and that positive aspects of work (i.e., resources, task autonomy, and support) would be related to greater helping behavior for those with higher SPS (Vander Elst et al., 2019). Also, the study examined specific aspects of SPS such as EOE, AES, and LST. Survey data from 1019 employees, with average tenure of about 20 years in their current role, showed that individuals reporting greater EOE and LST had greater emotional exhaustion with increasing job demands. On the positive side, LST was positively associated with helping behavior for high SPS persons with more job resources.

These findings are in line with other research showing strong correlations between SPS and overall work stress, as well as work displeasure and need for recovery from work, both characteristics of the *burnout* stage of the stress process (Evers et al., 2008). However, it seems that these effects are moderated by the specific SPS facets. Only two facets of SPS (EOE and LST) were related to work stress, work displeasure and need for recovery. The findings are intuitive, since the third facet, AES, is characterized by the more positive aspects of SPS. Despite the fact that work stress has been linked to negative affectivity, the relationship between SPS and work stress was not exclusively determined by the impact of negative affectivity.

Also, on the positive side, for a trait associated with vigilance and caution in new situations (Aron & Aron, 1997), a higher-than-average proportion of expatriate workers are highly sensitive (Andresen et al., 2018). Workplace superiors value the SPS traits of empathy and conscientiousness, and thus prefer to send highly sensitive workers abroad (versus those less sensitive). In a study of expatriate employees, Andresen et al. (2018) found that 26% of expats were highly sensitive, compared to the 20%–31% ratio in the general population (Aron & Aron, 1997; Lionetti et al., 2018). Highly sensitive individuals, especially when sent on international assignments, can experience significant stress related to meeting new people, being in unfamiliar situations, and forming new social bonds. It is social stress and uncertainty about novel situations, rather than simply being highly sensitive, that is related to higher turnover (or intention to quit employment) among expatriate HSPs. The authors suggest that one-on-one mentoring or (single) coaching at the company (and also, if possible during the assignments), could mitigate expatriates' stress.

3.4 Social outcomes

3.4.1 Relationships

SPS also affects marital relationships, both for better and for worse (Haase et al., 2013). For example, a neuroimaging study by Acevedo et al. (2014) showed that highly sensitive individuals displayed greater empathy-related neural activity in response to emotionally evocative (happy and sad) face images of their spouse (versus stranger and neutral images). Also, highly (versus less) sensitive participants showed stronger activity in the ventral tegmental area (VTA), a key reward region of the brain, in response to viewing happy face images of the partner.

High sensitivity also seems to be related to lower levels of marital quality, which varies as a function of childhood environment and current levels of life stress. For example, with a modest sample of 39 individuals, researchers found that high sensitivity was related to lower levels of self-reported romantic love, friendship-based love, and relationship happiness stability (Acevedo, Aron, & Aron, 2018). However, when the effects of positive childhood were accounted for, high SPS became positively associated with romantic love. Moreover, when considering the effects of past 6-month life stress (excluding relationship stress), SPS became positively correlated with friendship-based love and promotion focus (for growth and approach) in the relationship.

Other research from a 13-year longitudinal study of middle-aged and older adults in long-term marriages found that the short allele of the 5-HTTLPR polymorphism of the serotonin transporter gene, a marker of both SPS (Licht et al., 2011) and of differential susceptibility (Van Ijzendoorn, Belsky, & Bakermans-Kranenburg, 2012), was associated with amplified sensitivity to both negative and positive emotional behavior in marriages (Haase et al., 2013). Specifically, individuals with two short alleles of 5-HTTLPR (s/s), experienced less marital satisfaction over time in a climate characterized by high levels of negative emotional behavior and low levels of positive emotional behavior, and more marital satisfaction over time in a climate characterized by low levels of negative emotional behavior and high levels of positive emotional behavior.

In sum, these results are consistent with theoretical models of ES and vantage sensitivity, which propose that individuals with greater "sensitivity" to the environment express the trait differently depending on both rearing and current contextual factors. Thus, highly sensitive individuals with low-quality childhoods may especially benefit from family-based skills training, supportive environments, and techniques focused on enhancing emotion regulation, communication, and reactions to stress. Additionally, this information may be useful to couples (with one or two highly sensitive partners) undergoing challenging times so they do not simply jump ship but realize that current life stress may be bringing out the "worst" features of the SPS trait. However, as the stressors pass, hopefully with time and modifications to stressful circumstances, the more positive side of the trait will prevail. More detail is provided in the chapter on clinical issues related to SPS.

3.4.2 Communication

Communication is also a critical aspect of social relationships, and within a dialogue, listening has been linked to well-being (Lun, Kesebir, & Oishi, 2008; Sarason, Sarason, & Gurung, 1997). Good listening makes the speaker feel understood, and in turn, feeling understood is a proximal cause of well-being. Itzchakov, DeMarree, Kluger, and Turjeman-Levi (2018) found that empathic listening was associated with decreased social anxiety for the individual speaking. Listening has been linked to empathy (Dolev & Kluger, 2011) and individuals high in SPS experience more empathy than do less-sensitive individuals (Acevedo et al., 2014; Aron & Aron, 1997). However, research suggests that overstimulating environments can overwhelm HSPs, keeping them from being able to focus on their listener. Recent findings showed that distraction turned high SPS individuals into especially poor listeners (Turjeman-Levi, 2016). In turn, when the speakers became aware of the listeners' distractedness, the listeners' social anxiety increased.

Highly sensitive individuals also report feeling more nervous when being observed (Aron & Aron, 1997). Recent evidence expanding on findings suggests that it may be the serotonin transporter (5-HTTLPR) polymorphism associated with SPS which interacts with stressful conditions, resulting in dysregulated physiological arousal, possibly modulated by cortisol levels. For example, Alexander et al. (2010) genotyped healthy male adults with no history of psychopathology and asked them to engage in a public speaking task, which has been shown to induce stress. Saliva cortisol levels were obtained at six time points, prior to the stressor and during an extended recovery period. Findings showed that individuals homozygous for the 5-HTTLPR short allele associated with SPS, compared to all other groups, showed elevated cortisol reactivity in response to the public speaking task, but only if they reported a high number of severe stressful life events. Thus, current life stress moderates the impact that public speaking (or any stressor) has on individuals, but the impact of stressful situations is generally amplified for those that are highly sensitive, and especially when they are already experiencing life stress.

4 Conclusion

Although it is a normal variation in temperament, high SPS can predispose individuals to physical and psychological issues (Belsky, Bakermans-Kranenburg, & van Ijzendoorn, 2007; Belsky & Pluess, 2009; Boyce et al., 1995; Fox et al., 2007; Gannon et al., 1989) under harsh conditions. For example, SPS has been linked with poor social outcomes including negative parent-child interactions in childhood (Degnan & Fox, 2007), difficulties with peers in youth (Nocentini et al., 2018), and lower relationship quality in adult marriages (Acevedo, Aron, & Aron, 2018). SPS has also been associated with more self-reported physical health problems and pain (Benham, 2006; DeGangi, 2000; Engel-Yeger & Dunn, 2011; Liss et al., 2005), lower work and life satisfaction (Andresen et al., 2018; Evers et al., 2008), difficulty managing stress (Brindle et al., 2015), and mental health disorders such as anxiety, depression, internalizing problems, emotion regulation difficulties, and panic disorder (Bakker & Moulding, 2012; Brindle et al., 2015; Liss et al., 2005, 2008; Neal et al., 2002).

On the positive side, high SPS individuals reporting positive (versus harsh, poor, or abusive) childhood environments also report greater life satisfaction (Booth et al., 2015) and higher relationship quality as adults (Acevedo, Aron, & Aron, 2018). These effects are also shown when accounting for high levels of current life stress, suggesting that current environmental conditions also have a stronger impact on how highly sensitive individuals experience the quality of their lives and relationships. Also, on the positive side, research suggests that high sensitivity is associated with more creativity and better listening skills (which occur in situations of low stimulation). Some propose that the neuronal plasticity/flexibility, which allows the SPS trait to express itself differentially according to the environment, may be associated with genetic polymorphisms in serotonergic and dopaminergic systems (Chen et al., 2011; Licht et al., 2011). Herein we provide a glimpse, since the genetic and neural basis of SPS will be discussed in detail in Chapter 5.

Capitalizing on the flexibility of the SPS trait, researchers are increasingly using interventions at the early stages of development to take advantage of the neural plasticity inherent in SPS, not only to inoculate children against negative mental health outcomes, but also to help them flourish (Nocentini et al., 2018; Pluess & Boniwell, 2015). Future directions should involve further investigation of the long-term outcomes of interventions into adulthood. Also, it would be beneficial to understand the environmental factors associated with positive health outcomes among highly sensitive individuals in order to create environments and design interventions to harness the best of high SPS individuals. Similarly, further investigation of biological mechanisms may elucidate pathways, such as the HPA axis (Alexander et al., 2010; Blunt Bugental et al., 2003; Hastings et al., 2014; Nachmias et al., 1996), which may be targeted in future medical or health treatments.

In sum, research on the health and social outcomes of SPS suggests that in impoverished or negative environments, highly sensitive individuals show a predisposition to a variety of health and psychological symptoms ranging from heightened pain to severe depression. However, the positive aspects of the trait such as good listening, greater love, empathy, creativity, and aesthetic sensibilities tend to be expressed to a greater degree among the highly sensitive under positive, nurturing, and relatively calm environmental conditions. Although it is beneficial to foster positive environments to enable highly sensitive individuals to thrive, when seemingly small (or even great) changes are made via interventions, policies, and awareness, it's likely to promote the greater good for all.

References

Acevedo, B., Aron, E., Pospos, S., & Jessen, D. (2018). The functional highly sensitive brain: A review of the brain circuits underlying sensory processing sensitivity and seemingly related disorders. *Philosophical Transactions of the Royal Society B Biological Sciences*, *373*(1744). https://doi.org/10.1098/rstb.2017.0161.

Acevedo, B. P., Aron, E. N., & Aron, A. (2018). *Novel perspectives on sensory processing sensitivity*. San Francisco, CA, USA: Association for Psychological Science Convention.

Acevedo, B. P., Aron, E. N., Aron, A., Sangster, M. D., Collins, N., & Brown, L. L. (2014). The highly sensitive brain: An fMRI study of sensory processing sensitivity and response to others' emotions. *Brain and Behavior*, *4*(4), 580–594.

Acevedo, B. P., Jagiellowicz, J., Aron, E., Marhenke, R., & Aron, A. (2017). Sensory processing sensitivity and childhood quality's effects on neural responses to emotional stimuli. *Clinical Neuropsychiatry*, *14*(6), 359–373.

Ahadi, & Basharpoor. (2010). Relationship between sensory processing sensitivity, personality dimensions and mental health. *Journal of Applied Sciences*, *10*, 570–574.

Akinola, M., & Mendes, W. B. (2008). The dark side of creativity: Biological vulnerability and negative emotions lead to greater artistic creativity. *Personality and Social Psychology Bulletin*, *34*(2), 167–1686. https://doi.org/10.1177/0146167208323933.

Alexander, N., Osinsky, R., Schmitz, A., Mueller, E., Kuepper, Y., & Hennig, J. (2010). The BDNF Val66Met polymorphism affects HPA-axis reactivity to acute stress. *Psychoneuroendocrinology, 35*(6), 949–953. https://doi.org/10.1016/j.psyneuen.2009.12.008.

American Psychiatric Association. (2013). *Diagnostic and statistical manual of mental disorders* (5th ed.). Arlington, VA: American Psychiatric Publishing.

Andresen, M., Goldmann, P., & Volodina, A. (2018). Do overwhelmed expatriates intend to leave? The effects of sensory processing sensitivity, stress, and social capital on expatriates' turnover intention. *European Management Review, 15*, 315–328. https://doi.org/10.1111/emre.12120.

Aron, E. N., & Aron, A. (1997). Sensory-processing sensitivity and its relation to introversion and emotionality. *Journal of Personality and Social Psychology, 73*(2), 345–368. https://doi.org/10.1037/0022-3514.73.2.345.

Aron, E. N., Aron, A., & Davies, K. M. (2005). Adult shyness: The interaction of temperamental sensitivity and an adverse childhood environment. *Personality and Social Psychology Bulletin, 31*(2), 181–197. https://doi.org/10.1177/0146167204271419.

Aron, E. N., Aron, A., & Jagiellowicz, J. (2012). Sensory processing sensitivity: A review in the light of the evolution of biological responsivity. *Personality and Social Psychology Review, 16*(3), 262–282. https://doi.org/10.1177/1088868311434213.

Bakker, K., & Moulding, R. (2012). Sensory-processing sensitivity, dispositional mindfulness and negative psychological symptoms. *Personality and Individual Differences, 53*(3), 341–346. https://doi.org/10.1016/j.paid.2012.04.006.

Belsky, J. (1997). Variation in susceptibility to rearing influence: An evolutionary argument. *Psychological Inquiry, 8*, 182–186.

Belsky, J. (2005). Differential susceptibility to rearing influence: An evolutionary hypothesis and some evidence. In B. Ellis & D. Bjorklund (Eds.), *Origins of the social mind: Evolutionary psychology and child development* (pp. 139–163). New York: Guilford.

Belsky, J., Bakermans-Kranenburg, M. J., & van Ijzendoorn, M. H. (2007). For better and for worse differential susceptibility to environmental influences. *Current Directions in Psychological Science, 16*(6), 300–304.

Belsky, J., Hsieh, K. H., & Crnic, K. (1998). Mothering, fathering, and infant negativity as antecedents of boys' externalizing problems and inhibition at age 3 years: Differential susceptibility to rearing experience? *Development and Psychopathology, 10*(2), 301–319.

Belsky, J., & Pluess, M. (2009). Beyond diathesis stress: Differential susceptibility to environmental influences. *Psychological Bulletin, 135*(6), 885–908. https://doi.org/10.1037/a0017376.

Benham, G. (2006). The highly sensitive person: Stress and physical symptom reports. *Personality and Individual Differences, 40*(7), 1433–1440. https://doi.org/10.1016/j.paid.2005.11.021.

Biederman, J., Rosenbaum, J. F., Hirshfeld, D. R., Faraone, S.V., Bolduc, E. A., & Gersten, M. (1990). Psychiatric correlates of behavioral inhibition in young children of parents with and without psychiatric disorders. *Archives of General Psychiatry, 47*, 21–26. https://doi.org/10.1001/archpsyc.1990.01810130023004.

Blunt Bugental, D., Martorell, G. A., & Barazza, V. (2003). The hormonal costs of subtle forms of infant maltreatment. *Hormones and Behavior, 43*, 237–244. https://doi.org/10.1016/S0018-506X(02)00008-9.

Boeke, E. A., Moscarello, J. M., LeDoux, J. E., Phelps, E. A., & Hartley, C. A. (2017). Active avoidance: Neural mechanisms and attenuation of pavlovian conditioned responding. *Journal of Neuroscience, 37*(18), 4808–4818. https://doi.org/10.1523/JNEUROSCI.3261-16.2017.

Boniwell, I., & Ryan, L. (2009). *SPARK Resilience: A teacher's guide*. London, UK: University of East London.

Booth, C., Standage, H., & Fox, E. (2015). Sensory-processing sensitivity moderates the association between childhood experiences and adult life satisfaction. *Personality and Individual Differences, 87*, 24–29. https://doi.org/10.1016/j.paid.2015.07.020.

Boterberg, S., & Warreyn, P. (2016). Making sense of it all: The impact of sensory processing sensitivity on daily functioning of children. *Personality and Individual Differences, 92,* 80–86. https://doi.org/10.1016/j.paid.2015.12.022.

Boyce, W. T., Chesney, M., Alkon, A., Tschann, J. M., Adam, S., Chesterman, B., … Wara, D. (1995). Psychobiologic reactivity to stress and childhood respiratory illnesses: Results of two prospective studies. *Psychosomatic Medicine, 57,* 411–422.

Boyce, W. T., & Ellis, B. J. (2005). Biological sensitivity to context. I. An evolutionary-developmental theory of the origins and functions of stress reactivity. *Development and Psychopathology, 17*(2), 271–301. https://doi.org/10.1017/S0954579405050145.

Branjerdporn, G., Meredith, P., Strong, J., & Green, M. (2019). Sensory sensitivity and its relationship with adult attachment and parenting styles. *PLoS One, 14*(1), e0209555. https://doi.org/10.1371/journal.pone.0209555.

Bridges, D., & Schendan, H. E. (2019). Sensitive individuals are more creative. *Personality and Individual Differences, 142,* 186–195. https://doi.org/10.1016/j.paid.2018.09.015.

Brindle, K., Moulding, R., Bakker, K., & Nedeljkovic, M. (2015). Is the relationship between sensory-processing sensitivity and negative affect mediated by emotional regulation? *Australian Journal of Psychology, 67*(4), 214–221. https://doi.org/10.1111/ajpy.12084.

Carter, C. S., Grippo, A. J., Pournajafi-Nazarloo, H., Ruscio, M. G., & Porges, S. W. (2008). Oxytocin, vasopressin and sociality. *Progress in Brain Research, 170,* 331–336. https://doi.org/10.1016/S0079-6123(08)00427-5.

Caspi, A., Sugden, K., Moffitt, T. E., Taylor, A., Craig, I. W., Harrington, H., … Poulton, R. (2003). Influence of life stress on depression: Moderation by a polymorphism in the 5-HTT Gene. *Science, 301*(5631), 386–389. https://doi.org/10.1126/science.1083968.

Chen, C., Chen, C., Moyzis, R., Sten, H., He, Q., Li, H., … Dong, Q. (2011). Contributions of dopamine-related genes and environmental factors to highly sensitive personality: A multi-step neuronal system-level approach. *PLoS One, 6*(7), e21636. https://doi.org/10.1371/journal.pone.0021636.

Chronis-Tuscano, A., Degnan, K. A., Pine, D. S., Perez-Edgar, K., Henderson, H. A., Diaz, Y., … Fox, N. A. (2009). Stable early maternal report of behavioral inhibition predicts lifetime social anxiety disorder in adolescence. *Journal of the American Academy of Child and Adolescent Psychiatry, 48,* 928–935.

Coplan, R. J., Wilson, J., Frohlick, S. L., & Zelenski, J. (2006). A person-oriented analysis of behavioral inhibition and behavioral activation in children. *Personality and Individual Differences, 41*(5), 917–927. https://doi.org/10.1016/j.paid.2006.02.019.

Costa, P. T., & McCrae, R. P. (1992). Four ways five factors are basic. *Personality and Individual Differences, 13*(6), 653–665. https://doi.org/10.1016/0191-8869(92)90236-I.

Creed, P. A., Muller, J., & Machin, M. A. (2001). The role of satisfaction with occupational status, neuroticism, financial strain and categories of experience in predicting mental health in the unemployed. *Personality and Individual Differences, 30*(3), 435–447. https://doi.org/10.1016/S0191-8869(00)00035-0.

DeGangi. (2000). Prediction of childhood problems at three years in children experiencing disorders of regulation during infancy. *Infant Mental Health Journal, 21*(3), 156–175. https://doi.org/10.1002/1097-0355(200007)21:3<156::AID-IMHJ2>3.0.CO;2-D.

Degnan, K. A., & Fox, N. A. (2007). Behavioral inhibition and anxiety disorders: Multiple levels of a resilience process. *Development and Psychopathology, 19*(3), 729–746. https://doi.org/10.1017/S0954579407000363.

Degnan, K. A., Henderson, H. A., Fox, N. A., & Rubin, K. H. (2008). Predicting social wariness in middle childhood: The moderating roles of child care history, maternal personality and maternal behavior. *Social Development, 17*(3), 471–487. https://doi.org/10.1111/j.1467-9507.2007.00437.x.

Derryberry, & Rothbart. (1997). Reactive and effortful processes in the organization of temperament. *Development and Psychopathology, 9*(4), 633–652. https://doi.org/10.1017/S0954579497001375.

Dolev, A., & Kluger, A. N. (2011). *The effects of supervisors listening on subordinates well-being*. [Chair]. In Kluger, A. N. (Ed.), *Symposium presented at the 1st Israel organizational behavior conference, Tel Aviv, Israel.*

Eisenberg, D., Downs, M. F., Golberstein, E., & Zivin, K. (2009). Stigma and help seeking for mental health among college students. *Medical Care Research and Review, 66*(5), 522–541. https://doi.org/10.1177/1077558709335173.

Eisenberg, N., Damon, W., & Lerner, R. M. (2006). *Handbook of child psychology: Social, emotional, and personality development* (6th ed.). Hoboken, NJ, US: John Wiley & Sons Inc.

Eisenberg, N., Guthrie, I. K., Cumberland, A., Murphy, B. C., Shepard, S. A., Zhou, Q., & Gustavo, C. (2002). *Prosocial development in early adulthood: A longitudinal study* (p. 298). Faculty Publications, Department of Psychology.

Engel-Yeger, B., & Dunn, W. (2011). Relationship between pain catastrophizing level and sensory processing patterns in typical adults. *American Journal of Occupational Therapy, 65*, e1–e10.

Evans, D. E., & Rothbart, M. K. (2008). Temperamental sensitivity: Two constructs or one? *Personality and Individual Differences, 44*, 108–118. https://doi.org/10.1016/j.paid.2007.07.016.

Evers, A., Rasche, J., & Schabracq, M. J. (2008). High sensory-processing sensitivity at work. *International Journal of Stress Management, 15*(2), 189–198. https://doi.org/10.1037/1072-5245.15.2.189.

Feldman, R., Greenbaum, C.W., & Yirmiya, N. (1999). Mother-infant affect synchrony as an antecedent of the emergence of self-control. *Developmental Psychology, 35*(1), 223–231. https://doi.org/10.1037/0012-1649.35.1.223.

First, M. B., Gibbon, M., Spitzer, R. L., Williams, J. B. W., & Benjamin, L. S. (1997). *Structured clinical interview for DSM IV axis I and II disorders*. Washington: APA.

Fox, M. A., Sanes, J. R., Borza, D. B., Eswarakumar, V. P., Fässler, R., Hudson, B. G., … Umemori, H. (2007). Distinct target-derived signals organize formation, maturation, and maintenance of motor nerve terminals. *Cell, 129*(1), 179–193. https://doi.org/10.1016/j.cell.2007.02.035.

Fox, N. A., Henderson, H. A., Marshall, P. J., Nichols, K. E., & Ghera, M. M. (2005). Behavioral inhibition: Linking biology and behavior within a developmental framework. *Annual Review of Psychology, 56*, 235–262. https://doi.org/10.1146/annurev.psych.55.090902.141532.

Fox, N.A., Henderson, H.A., Rubin, K. H., Calkins, S. D., & Schmidt, L.A. (2001). Continuity and discontinuity of behavioral inhibition and exuberance: Psychophysiological and behavioral influences across the first four years of life. *Child Development, 72*(1), 1–21. https://doi.org/10.1111/1467-8624.00262.

Furnham, A., & Cheng, H. (1999). Personality as predictor of mental health and happiness in the East and West. *Personality and Individual Differences, 27*(3), 395–403. https://doi.org/10.1016/S0191-8869(98)00250-5.

Gannon, L., Banks, J., Shelton, D., & Luchetta, T. (1989). The mediating effects of psychophysiological reactivity and recovery on the relationship between environmental stress and illness. *Journal of Psychosomatic Research, 33*(2), 167–175. https://doi.org/10.1016/0022-3999(89)90044-5.

Gladstone, G. L., Parker, G. B., Mitchell, P. B., Wilhelm, K. A., & Malhi, G. S. (2005). Relationship between self-reported childhood behavioral inhibition and lifetime anxiety disorders in a clinical sample. *Depression and Anxiety, 22*(3), 103–113. https://doi.org/10.1002/da.20082.

Goetz, T. E., & Dweck, C. S. (1980). Learned helplessness in social situations. *Journal of Personality and Social Psychology, 39*(2), 246–255. https://doi.org/10.1037/0022-3514.39.2.246.

Gottman, J. M., Katz, L. F., & Hooven, C. (1997). *Meta-emotion: How families communicate emotionally*. Hillsdale, NJ, US: Lawrence Erlbaum Associates, Inc. https://doi.org/10.4324/9780203763568.

Gunnar, M. R. (1994). Psychoendocrine studies of temperament and stress in early childhood: Expanding current models. In J. E. Bates & T. D. Wachs (Eds.), *Temperament: Individual differences at the interface of biology and behavior.* APA. https://doi.org/10.1037/10149-006.

Haase, C. M., Saslow, L. R., Bloch, L., Saturn, S. R., Casey, J. J., Seider, B. H., … Levenson, R. W. (2013). The 5-HTTLPR polymorphism in the serotonin transporter gene moderates the association between emotional behavior and changes in marital satisfaction over time. *Emotion, 13*(6), 1068–1079. https://doi.org/10.1037/a0033761.

Hane, A. A., Cheah, C., Rubin, K. H., & Fox, N. A. (2008). The role of maternal behavior in the relation between shyness and social withdrawal in early childhood and social withdrawal in middle childhood. *Social Development, 17*, 795–811. https://doi.org/10.1111/j.1467-9507.2008.00481.x.

Hanes, A. (2016). *The impact of social anxiety and sensory processing sensitivity on quality of life.* ProQuest. 10134131.

Hankin, B. L., Nederhof, E., Oppenheimer, C. W., Jenness, J., Young, J. F., Abela, J. R. Z., … Oldehinkel, A. J. (2011). Differential susceptibility in youth: Evidence that 5-HTTLPR x positive parenting is associated with positive affect 'for better and worse'. *Translational Psychiatry, 1*, e44. https://doi.org/10.1038/tp.2011.44.

Hartley, C. A., Gorun, A., Reddan, M. C., Ramirez, F., & Phelps, E. A. (2014). Stressor controllability modulates fear extinction in humans. *Neurobiology of Learning and Memory, 113*, 149–156. https://doi.org/10.1016/j.nlm.2013.12.003.

Hartman, S., & Belsky, J. (2018). Prenatal stress and enhanced developmental plasticity. *Journal of Neural Transmission (Vienna), 125*, 1759–1779. https://doi.org/10.1007/s00702-018-1926-9.

Hastings, P. D., Klimes-Dougan, B., Kendziora, K. T., Brand, A., & Zahn-Waxler, C. (2014). Regulating sadness and fear from outside and within: mothers' emotion socialization and adolescents' parasympathetic regulation predict the development of internalizing difficulties. *Development and Psychopathology, 26*(4), 1369–1384. https://doi.org/10.1017/S0954579414001084.

Hastings, P. D., Zahn-Waxler, C., & Usher, B. A. (2007). Cardiovascular and affective responses to social stress in adolescents with internalizing and externalizing problems. *International Journal of Behavioral Development, 31*(1), 77–87. https://doi.org/10.1177/0165025407073575.

Hayward, C., Killen, J. D., Kraemer, H. C., & Taylor, B. (1998). Linking self-reported childhood behavioral inhibition to adolescent social phobia. *Journal of the American Academy of Child & Adolescent Psychiatry, 37*(12), 1308–1316. https://doi.org/10.1097/00004583-199812000-00015.

Henderson, H. A., Marshall, P. J., Fox, N. A., & Rubin, K. H. (2004). Psychophysiological and behavioral evidence for varying forms and functions of nonsocial behavior in preschoolers. *Child Development, 75*(1), 251–263. https://doi.org/10.1111/j.1467-8624.2004.00667.x.

Hirshfeld, D. R., Rosenbaum, J., Biederman, J., Bolduc, E. A., Faraone, S. V., Snidman, N., … Kagan, J. (1992). Stable behavioral inhibition and its association with anxiety disorder. *Journal of the American Academy of Child and Adolescent Psychiatry, 31*, 103–111. https://doi.org/10.1097/00004583-199201000-00016.

Hirshfeld-Becker, D. R., Biederman, J., Henin, A., Faraone, S. V., Davis, S., & Harrington, K. (2007). Behavioral inhibition in preschool children at risk is a specific predictor of middle childhood social anxiety: A five-year follow-up. *Journal of Developmental & Behavioral Pediatrics, 28*, 225–233. https://doi.org/10.1097/01.DBP.0000268559.34463.d0.

Hofmann, S. G., & Bitran, S. (2006). Sensory-processing sensitivity in social anxiety disorder: Relationship to harm avoidance and diagnostic subtypes. *Journal of Anxiety Disorders, 21*(7), 944–954. https://doi.org/10.1016/j.janxdis.2006.12.003.

Hudson, J. L., Comer, J. S., & Rapee, R. (2008). Parental responses to positive and negative emotions in anxious and nonanxious children. *Journal of Clinical Child and Adolescent Psychology, 37*, 303–313.

Hudson, J. L., & Rapee, R. M. (2002). Parent-child interactions in clinically anxious children and their siblings. *Journal of Clinical Child and Adolescent Psychology*, *31*(4), 548–555. https://doi.org/10.1080/15374410801955839.

Itzchakov, G., DeMarree, K. G., Kluger, A. N., & Turjeman-Levi, Y. (2018). The listener sets the tone: High-quality listening increases attitude clarity and behavior-intention consequences. *Personality and Social Psychology Bulletin*, *44*(5), 762–778. https://doi.org/10.1177/0146167217747874.

Jonsson, K., Grim, K., & Kjellgren, A. (2014). Do highly sensitive persons experience more nonordinary states of consciousness during sensory isolation? *Social Behavior and Personality*, *42*(9), 1495–1506. https://doi.org/10.2224/sbp.2014.42.9.1495.

Kagan, J. (1994). On the Nature of Emotion. *Monographs of the Society for Research in Child Development*, *59*(2–3), 7–24. https://doi.org/10.2307/1166136.

Kagan, J., & Moss, H. A. (1962). *Birth to maturity: A study in psychological development*. Hoboken, NJ, US: John Wiley & Sons Inc.

Kagan, J., Reznick, S., & Snidman, N. (1987). The physiology and psychology of behavioral inhibition in children. *Child Development*, *58*(6), 1459–1473. https://doi.org/10.2307/1130685.

Kagan, J., Snidman, N., Zentner, M., & Peterson, E. (1999). Infant temperament and anxious symptoms in school age children. *Development and Psychopathology*, *11*(2), 209–224. https://doi.org/10.1017/S0954579499002023.

Kaufman, J., Yang, B. Z., Dogulas-Palumberi, H., Houshyar, S., Lipschitz, D., Krystal, J. H., & Gelernter, J. (2004). Social supports and serotonin transporter gene moderate depression in maltreated children. *Proceedings of the National Academy of Sciences of the United States of America*, *101*(49), 17316–17321.

Kinnealey, M., & Fuiek, M. (1999). The relationship between sensory defensiveness, anxiety, depression and perception of pain in adults. *Occupational Therapy International*, *6*(3), 195–206. https://doi.org/10.1002/oti.97.

Klein Velderman, M., Bakermans-Kranenburg, M. J., Juffer, F., & van IJzendoorn, M. H. (2006). Effects of attachment-based interventions on maternal sensitivity and infant attachment: Differential susceptibility of highly reactive infants. *Journal of Family Psychology*, *20*, 266–274. https://doi.org/10.1037/0893-3200.20.2.266.

Kochanska, G., Aksan, N., & Joy, M. E. (2007). Children's fearfulness as a moderator of parenting in early socialization: Two longitudinal studies. *Developmental Psychology*, *43*(1), 222–237. https://doi.org/10.1037/0012-1649.43.1.222.

Kochanska, G., Kim, S., Barry, R. A., & Philibert, R. A. (2011). Children's genotypes interact with maternal responsive care in predicting children's competence: Diathesis–stress or differential susceptibility? *Development and Psychopathology*, *23*(2), 605–616. https://doi.org/10.1017/S0954579411000071.

Kumsta, R., Stevens, S., Brookes, K., Schlotz, W., Castle, J., Beckett, C., … Sonuga-Barke, E. (2010). 5HTT genotype moderates the influence of early institutional deprivation on emotional problems in adolescence: Evidence from the English and Romanian Adoptee (ERA) study. *Journal of Child Psychology and Psychiatry*, *51*(7), 755–762. https://doi.org/10.1111/j.1469-7610.2010.02249.x.

Licht, C. L., Mortensen, E. L., & Knudsen, G. M. (2011). Association between sensory processing sensitivity and the serotonin transporter polymorphism 5-HTTLPR short/short genotype. *Biological Psychiatry*, *69*(9), 152s–153s.

Lionetti, F., Aron, A., Aron, E. N., Burns, L., Jagiellowicz, J., & Pluess, M. (2018). Dandelions, tulips and orchids: Evidence for the existence of low-sensitive, medium-sensitive and high-sensitive individuals. *Translational Psychology*, *8*, 24.

Lionetti, F., Aron, E. N., Aron, A., Klein, D. N., & Pluess, M. (2019). Observer-rated environmental sensitivity moderates children's response to parenting quality in early childhood. *Developmental Psychology*, *55*, 2389–2402. https://doi.org/10.1037/dev0000795.

Liss, M., Mailloux, J., & Erchull, M. J. (2008). The relationships between sensory processing sensitivity, alexi-thymia, autism, depression, and anxiety. *Personality and Individual Differences*, *45*(3), 255–259.

Liss, M., Timmel, L., Baxley, K., & Killingsworth, P. (2005). Sensory processing sensitivity and its relation to parental bonding, anxiety, and depression. *Personality and Individual Differences*, *39*(8), 1429–1439. https://doi.org/10.1016/j.paid.2005.05.007.

Lu, L. (1994). University transition: Major and minor life stressors, personality characteristics and mental health. *Psychological Medicine*, *24*(1), 81–87. https://doi.org/10.1017/S0033291700026854.

Lu, L., & Shih, J. B. (1997). Personality and happiness: Is mental health a mediator? *Personality and Individual Differences*, *22*(2), 249–256. https://doi.org/10.1016/S0191-8869(96)00187-0.

Lun, J., Kesebir, S., & Oishi, S. (2008). On feeling understood and feeling well: The role of interdependence. *Journal of Research in Personality*, *42*(6), 1623–1628. https://doi.org/10.1016/j.jrp.2008.06.009.

Martindale, C. (1977). Creativity, consciousness, and cortical arousal. *Journal of Altered States of Consciosness*, *3*, 69–87.

Matheny, A. P., Wachs, T. D., Ludwig, J., & Phillips, K. (1995). Bringing order out of chaos: Psychometric characteristics of the Confusion, Hubbub, and Order Scale. *Journal of Applied Developmental Psychology*, *16*, 429–444. https://doi.org/10.1016/0193-3973(95)90028-4.

McDermott, J. M., & Fox, N. A. (2010). Exploring response-monitoring: Developmental differences and contributions to self-regulation. In R. H. Hoyle (Ed.), *Handbook of self-regulation and personality* (pp. 91–113). Oxford: Wiley-Blackwell Publishing.

Meredith, P. J., Bailey, K. J., Strong, J., & Rappel, G. (2016). Adult attachment, sensory processing, and distress in healthy adults. *The American Journal of Occupational Therapy*, *70*(1). https://doi.org/10.5014/ajot.2016.017376.

Meyer, B., Ajchenbrenner, M., & Bowles, D. P. (2005). Sensory sensitivity, attachment experiences, and rejection responses among adults with borderline and avoidant features. *Journal of Personality Disorders*, *19*(6), 641–658. https://doi.org/10.1521/pedi.2005.19.6.641.

Mick, M. A., & Telch, M. J. (1998). Social anxiety and history of behavioral inhibition in young adults. *Journal of Anxiety Disorders*, *12*(1), 1–20. https://doi.org/10.1016/S0887-6185(97)00046-7.

Mitchell, A. J., Chan, M., Bhatti, H., Grassi, L., Johansen, C., & Meader, N. (2011). Prevalence of depression, anxiety, and adjustment disorder in oncological, hematological, and palliative-care settings: A meta-analysis of 94 interview-based studies. *The Lancet Oncology*, *12*(2), 160–174. https://doi.org/10.1016/S1470-2045(11)70002-X.

Morrell, & Murray. (2003). Parenting and the development of conduct disorder and hyperactive symptoms in childhood: A prospective longitudinal study from 2 months to 8 years. *Journal of Child Psychology and Psychiatry*, *44*(4), 489–508. https://doi.org/10.1111/1469-7610.t01-1-00139.

Nachmias, M., Gunnar, M., Mangelsdorf, S., Parritz, R. H., & Buss, K. (1996). Behavioral inhibition and stress reactivity: The moderating role of attachment security. *Child Development*, *67*(2), 508–522.

Neal, J. A., Edelmann, R. J., & Glachan, M. (2002). Behavioral inhibition and symptoms of anxiety and depression: Is there a specific relationship with social phobia? *British Journal of Clinical Psychology*, *41*(Pt. 4), 361–374. https://doi.org/10.1111/j.1467-8624.1996.tb01748.x.

Nocentini, A., Menesini, E., & Pluess, M. (2018). The personality trait of environmental sensitivity predicts children's positive response to school-based anti-bullying intervention. *Clinical Psychological Science*, *6*(6), 848–859. https://doi.org/10.1177/2167702618782194.

Nolen-Hoeksema. (2000). The role of rumination in depressive disorders and mixed anxiety/depressive symptoms. *Journal of Abnormal Psychology*, *109*(3), 504–511. https://doi.org/10.1037/0021-843X.109.3.504.

Obradovic, J., Bush, N. R., Stamperdahl, J., Adler, N. E., & Boyce, W. T. (2010). Biological sensitivity to context: The interactive effects of stress reactivity and family adversity on socio-emotional behavior and school readiness. *Child Development, 81*(1), 270–289. https://doi.org/10.1111/j.1467-8624.2009.01394.x.

Park, S.Y., Belsky, J., Putnam, S., & Crnic, K. (1997). Infant emotionality, parenting, and 3-year inhibition: Exploring stability and lawful discontinuity in a male sample. *Developmental Psychology, 33*(2), 218–227.

Pfeiffer, B., & Kinnealey, M. (2003). Treatment of sensory defensiveness in adults. *Occupational Therapy International, 10*(3), 175–184. https://doi.org/10.1002/oti.184.

Pluess, M. (2015). Individual differences in environmental sensitivity. *Child Development Perspectives, 9*(3), 138–143. https://doi.org/10.1111/cdep.12120.

Pluess, M., & Belsky, J. (2010). Differential susceptibility to parenting and quality child care. *Developmental Psychology, 46*(2), 379–390. https://doi.org/10.1037/a0015203.

Pluess, M., & Belsky, J. (2013). Vantage sensitivity: Individual differences in response to positive experiences. *Psychological Bulletin, 139*(4), 901–916. https://doi.org/10.1037/a0030196.

Pluess, M., & Boniwell, I. (2015). Sensory-processing sensitivity predicts treatment response to a school-based depression prevention program: Evidence of vantage sensitivity. *Personality and Individual Differences, 82*, 40–45. https://doi.org/10.1016/j.paid.2015.03.011.

Porges, S. W., & Furman, S. A. (2011). The early development of the autonomic nervous system provides a neural platform for social behavior: A polyvagal perspective. *Infant and Child Development, 20*(1), 106–118. https://doi.org/10.1002/icd.688.

Posner, M. I., & Rothbart, M. K. (2000). Developing mechanisms of self-regulation. *Development and Psychopathology, 12*(3), 427–441. https://doi.org/10.1017/S0954579400003096.

Prior, M., Sanson, A., Smart, D., & Oberklaid, F. (2000). Does shy-inhibited temperament in childhood lead to anxiety problems in adolescence? *Journal of the American Academy of Child & Adolescent Psychiatry, 39*(4), 461–468. https://doi.org/10.1097/00004583-200004000-00015.

Reznick, J. S., Hegeman, I. M., Kaufman, E. R., Woods, S. W., & Jacobs, M. (1992). Retrospective and concurrent self-report of behavioral inhibition and their relation to adult mental health. *Development and Psychopathology, 4*(2), 301–321. https://doi.org/10.1017/S095457940000016X.

Rothbart, M. K., & Bates, J. E. (2006). Temperament. In W. Damon, R. Lerner, & N. Eisenberg (Eds.), *Handbook of child psychology: Vol. 3. Social, emotional, and personality development* (6th ed., pp. 99–166). New York: Wiley.

Rothbart, M. K., Ellis, L. K., Rueda, M. R., & Posner, M. I. (2003). Developing mechanisms of temperamental effortful control. *Journal of Personality, 71*(6), 1113–1143. https://doi.org/10.1111/1467-6494.7106009.

Rubin, A. M., Perse, E. M., & Powell, R. A. (1985). Loneliness, parasocial interaction, and local television news viewing. *Human Communication Research, 12*(2), 155–180. https://doi.org/10.1111/j.1468-2958.1985.tb00071.x.

Rubin, K. H., & Borwick, D. (1984). The communication skills of children who vary with regard to sociability. In H. Sypher & J. Applegates (Eds.), *Social cognition and communication*. Hillsdale, NJ: Lawrence Erlbaum Associates.

Rubin, K. H., Bukowski, W., & Parker, J. G. (1998). Peer interactions, relationships and groups. In *Handbook of Child Psychology* (pp. 619–700): Wiley. https://doi.org/10.1002/9780470147658.chpsy0310.

Rubin, K. H., Burgess, K. B., & Hastings, P. D. (2002). Stability and social-behavioral consequences of toddlers' inhibited temperament and parenting behaviors. *Child Development, 73*(2), 483–495. https://doi.org/10.1111/1467-8624.00419.

Rubin, K. H., Cheah, C. S. L., & Fox, N. (2001). Emotion regulation, parenting and display of social reticence in preschoolers. *Early Education and Development, 12*(1), 97–115. https://doi.org/10.1207/s15566935eed1201_6.

Rubin, K. H., Coplan, R. J., & Bowker, J. C. (2009). Social withdrawal in childhood. *Annual Review of Psychology, 60*, 141–171. https://doi.org/10.1146/annurev.psych.60.110707.163642.

Rubin, K. H., Hastings, P. D., Stewart, S. L., Henderson, H. A., & Chen, X. (1997). The consistency and concomitants of inhibition: Some of the children, all of the time. *Child Development, 68*(3), 467–483.

Rubin, K. H., Hymel, S., & Mills, R. S. L. (1989). Sociability and social withdrawal in childhood: Stability and outcomes. *Journal of Personality, 57*, 237–255. https://doi.org/10.1111/j.1467-6494.1989.tb00482.x.

Sarason, B. R., Sarason, I. G., & Gurung, R. A. R. (1997). *Close personal relationships and health outcomes: A key to the role of social support*. In *Handbook of personal relationships: Theory, research and interventions* (pp. 547–573): John Wiley & Sons.

Schriber, R. A., & Guyer, A. E. (2015). Adolescent neurobiological susceptibility to social context. *Developmental Cognitive Neuroscience, 19*, 1–18.

Schwartz, C. E., Snidman, N., & Kagan, J. (1999). Adolescent social anxiety as an outcome of inhibited temperament in childhood. *Journal of the American Academy of Child and Adolescent Psychiatry, 38*, 1008–1015. https://doi.org/10.1097/00004583-199908000-00017.

Shanahan, L., Copeland, W., Costello, E. J., & Angold, A. (2008). Specificity of putative psychosocial risk factors for psychiatric disorders in children and adolescents. *Journal of Child Psychology and Psychiatry, 49*, 34–42. https://doi.org/10.1111/j.1469-7610.2007.01822.x.

Slagt, M., Dubas, J. S., van Aken, M. A. G., Ellis, B. J., & Dekovic, M. (2017). Children's differential susceptibility to parenting: An experimental test of "for better and for worse". *Journal of Experimental Child Psychology, 154*, 78–97. https://doi.org/10.1016/j.jecp.2016.10.004.

Smolewska, K. A., McCabe, S. B., & Woody, E. Z. (2006). A psychometric evaluation of the Highly Sensitive Person Scale: The components of sensory-processing sensitivity and their relation to the BIS/BAS and "Big Five". *Personality and Individual Differences, 40*(6), 1269–1279. https://doi.org/10.1016/j.paid.2005.09.022.

Sobocko, K., & Zelenski, J. M. (2015). Trait sensory-processing sensitivity and subjective well-being: Distinctive associations for different aspects of sensitivity. *Personality and Individual Differences, 83*, 44–49. https://doi.org/10.1016/j.paid.2015.03.045.

Somerville, L. H. (2013). The teenage brain: Sensitivity to social evaluation. *Current Directions in Psychological Science, 22*, 121–127. https://doi.org/10.1177/0963721413476512.

Staub, E. (1979). *Positive social behavior and morality: Socialization and development*. (Vol. 2). New York: Academic Press.

Turjeman-Levi, Y. T. (2016). *When the sensitivity of one is the trouble of the other: High-sensitivity as a moderator of the effects of physical stimulation on listening*. Jerusalem: The Hebrew University of Jerusalem.

van Brakel, A. M., Muris, P., Bogels, S. M., & Thomassen, C. (2006). A multifactorial model for the etiology of anxiety in non-clinical adolescents: Main and interactive effects of behavioral inhibition, attachment and parental rearing. *Journal of Child and Family Studies, 15*, 568–578. https://doi.org/10.1007/s10826-006-9061-x.

Van Ijzendoorn, M. H., Belsky, J., & Bakermans-Kranenburg, M. J. (2012). Serotonin transporter genotype 5HTTLPR as a marker of differential susceptibility? A meta-analysis of child and adolescent gene-by-environment studies. *Translational Psychiatry, 2*. https://doi.org/10.1038/tp.2012.73.

Vander Elst, T., Sercu, M., Van den Broeck, A., Van Hoof, E., Baillien, E., & Godderis, L. (2019). Who is more susceptible to job stressors and resources? Sensory-processing sensitivity as a personal resource and vulnerability factor. *PLoS One, 14*(11), e0225103. https://doi.org/10.1371/journal.pone.0225103.

Wachs, S. (2012). Moral disengagement and emotional and social difficulties in bullying and cyberbullying: Differences by participant role. *Emotional and Behavioral Difficulties, 17*(3–4), 347–360. https://doi.org/10.1080/13632752.2012.704318.

Weyn, S., Van Leeuwen, K., Pluess, M., Lionetti, F., Greven, C., Goossens, L., … Bijettebier, P. (2019). Psychometric properties of the highly sensitive child scale across developmental stage, gender, and country. *Current Psychology*, 1–17. https://doi.org/10.1007/s12144-019-00254-5.

White, L. K., McDermott, J. M., Degnan, K. A., Henderson, H. A., & Fox, N. A. (2011). Behavioral inhibition and anxiety: The moderating roles of inhibitory control and attention shifting. *Journal of Abnormal Child Psychology, 39*(5), 735–747. https://doi.org/10.1007/s10802-011-9490-x.

Whiteside-Mansell, L., Bradley, R., Tresch Owen, M., Randolph, S. M., & Cauce, A. M. (2003). Parenting and children's behavior at 36 Months: Equivalence between African American and European American mother—Child dyads. *Parenting, 3*, 197–234. https://doi.org/10.1207/S15327922PAR0303_02.

Wood, J. J., McLeod, B. D., Sigman, M., Hwang, W. C., & Chu, B. C. (2003). Parenting and childhood anxiety: theory, empirical findings, and future directions. *Journal of Child Psychology and Psychiatry, 44*(1), 134–151. https://doi.org/10.1111/1469-7610.00106.

Wyller, H. B., Wyller, H.V. B., Crane, C., & Gjelsvik, B. (2018). The relationship between sensory processing sensitivity and psychological distress: A model of underpinning mechanisms and an analysis of therapeutic possibilities. *Scandinavian Journal of Psychology, 4*(15), 1–15.

CHAPTER 5

Etiology of sensory processing sensitivity: Neurobiology, genes, and evolution

Flavia V. Naumann[a], Bianca P. Acevedo[b], Jadzia Jagiellowicz[c], Corina U. Greven[a,d,e], Judith R. Homberg[a]
[a]Department of Cognitive Neuroscience, Donders Institute for Brain, Cognition and Behaviour, Radboud University Medical Center, Nijmegen, The Netherlands
[b]Neuroscience Research Institute, University of California, Santa Barbara, CA, United States
[c]Department of Psychology, Stony Brook University, Stony Brook, NY, United States
[d]Karakter Child and Adolescent Psychiatry University Center, Nijmegen, The Netherlands
[e]Social, Genetic & Developmental Psychiatry Centre, Institute of Psychiatry, Psychology & Neuroscience, King's College London, London, United Kingdom

Contents

1 Evolutionary basis for sensory processing sensitivity 109
2 Genes underlying SPS 110
2.1 Epigenetics 111
2.2 Serotonin 112
2.3 Dopamine 114
2.4 Other genes 116
3 The neural basis of SPS 118
3.1 Neurobiology of SPS: Evidence from human studies 119
3.2 Neurobiology of SPS: Evidence from animal studies 122
4 Peripheral sensitivities 125
4.1 Hypothalamus-pituitary-adrenal axis 125
4.2 Glucose metabolism and sensations 126
5 Summary and future perspectives 126
References 128

1 Evolutionary basis for sensory processing sensitivity

As discussed in depth in previous chapters of this book, sensory processing sensitivity (SPS) is thought to have evolved as one out of two evolutionary strategies in response to the environment. One theory suggests that individual differences in susceptibility to the environment and other cues depend on greater or lesser adaptation and plasticity, leading to more or less responsivity to stimuli (Belsky, Hsieh, & Crnic, 1998; Belsky & Pluess, 2009). Being more responsive

https://doi.org/10.1016/B978-0-12-818251-2.00005-9

to their environments, highly sensitive persons (HSPs) have an enhanced awareness of and emotional reactivity to opportunities (e.g., food, mates, and alliances) and also threats (e.g., predators, loss of status, competitors). HSPs process information deeply, engaging in complex thinking and making connections across themes. As such, highly sensitive individuals are ready to respond to emerging situations, such as by behaving cooperatively, responding to another's needs and also by perceiving threats (Acevedo et al., 2014; Acevedo, Jagiellowicz, Aron, Marhenke, & Aron, 2017). This survival strategy is thought to be effective as long as the benefits of increased sensitivity outweigh the costs, such as increased cognitive and metabolic demand (Wolf, van Doorn, & Weissing, 2008).

Other evidence of the evolutionary basis of SPS comes from animal studies, which suggest that it appears in over 100 different species, including varieties ranging from insects to fish (Schjolden & Winberg, 2007), and mammals (Sih, Bell, & Johnson, 2004), including goats (Lyons, Price, & Moberg, 1988), rats (Coppens, de Boer, & Koolhaas, 2010), primates (Suomi, 2006), and pigs (Hessing, Hagelsø, Schouten, Wiepkema, & van Beek, 1994). SPS appears across many species, thus it has been suggested that there must be an evolutionary advantage for it (Acevedo et al., 2014). Traits are namely thought to be preserved by natural selection when they contribute to individual- and species-level fitness, leading to an evolutionary advantage.

Other theories, as discussed previously in this book, suggest that whether an individual is highly sensitive and to what degree, depend on plasticity factors, which are mainly determined by genetic variants and environmental factors influencing the development of endophenotypic attributes (Belsky, 1997; Belsky & Pluess, 2009).

A large body of literature has shown that synergistic interactions between genes and the environment (e.g., parental conditions) contribute to the development of an individual's phenotype. Therefore, SPS and the degree of sensitivity associated with it are largely determined by genetic, childhood, and environmental factors. This has been covered at length in previous chapters. Next, we turn to the genetic factors thought to underlie SPS.

2 Genes underlying SPS

At least one twin study suggests that 47% of interindividual differences in SPS are due to genetic influences, with the remaining 53% largely being affected by the environment (Assary, Zavos, Krapohl, & Pluess, 2019). Thus, approximately half of the etiology of SPS is influenced by environmental factors. According to one view, the immediate expression of SPS may be more affected by "nonshared environments" than so-called shared environments (shared environments are those that make family members more similar to one another). Nonshared environments are defined as any environment that contribute to differences between family members. Examples of nonshared factors include illness, accidents, peer influences, romantic partners, and different perceptions of the (same) environment. Differential influences of the environment on highly sensitive individuals may be explained by the fact that positive, negative, and also neutral cues are noticed by high SPS individuals to a finer degree, with corresponding responsiveness. This ultimately affects development and how the SPS trait expresses itself phenotypically (Bridges & Schendan, 2018a, 2018b).

The following section discusses genes associated with SPS, coming both from direct and indirect evidence. We discuss research on genes with more speculative connections to SPS, outlining possibilities for future research.

2.1 Epigenetics

How SPS-related genes get expressed as well as the degree of expression may depend on the specific environmental conditions an individual is exposed to, including the prenatal and postnatal environments (Forssman et al., 2014; Pluess, 2015; Pluess & Belsky, 2011; Pluess et al., 2011; van Ijzendoorn & Bakermans-Kranenburg, 2015). Both interactions most likely involve the regulation of gene expression through environmentally induced epigenetic changes (Meaney, 2010; Szyf, 2015).

For example, a study of prairie voles suggested that prenatal risk for anxiety is transmitted by means of rearing environment, physiological reactivity (corticosterone reactivity to a stressor), and neuroendocrine function measured by oxytocin- and vasopressin-receptor binding (Hartman, Freeman, Bales, & Belsky, 2018). The study exposed a group of pregnant voles to an unfamiliar aggressive female vole, thus triggering a stress response in the pregnant voles. The other group did not experience these stressful situations during pregnancy. After birth, the newborns were placed with adoptive parents, which were either good caregivers or neglected the newborns. Testing for anxious behavior in adulthood, stressed rodent prairie voles crossfostered to high-quality rearing in infancy, displayed the least anxiety-like behavior and the least corticosterone reactivity of all animals in response to social isolation and a forced swim test (placement in a bin with water from which the animal cannot escape while active swimming or passive floating is measured). In comparison, stressed voles crossfostered to low-quality care in infancy displayed the most anxiety-like behavior and the most corticosterone reactivity of all animals when exposed to the same stress tests. Additionally, while neither prenatal stress nor rearing condition affected oxytocin-receptor binding, high-quality rearing predicted the highest vasopressin-1a receptor binding in the amygdala, an area classically known for its involvement in emotion processing. Vasopressin-1a receptor density in the amygdala is related to social and anxiety behaviors (Carter, Grippo, Pournajafi-Nazarloo, Ruscio, & Porges, 2008). For voles in the control group, none of the outcomes were related to the rearing condition. Results indicate that prenatal stress induces greater environmental sensitivity, making it both a risk and an opportunity factor.

2.2 Serotonin

To date, the serotonin transporter (5-HTT) linked polymorphic region (5-HTTLPR) short (s) allelic variant, which is associated with reduced transcription of the serotonin transporter gene compared to the long (l) allelic variant has been identified in association with SPS. As a neuromodulator, serotonin performs many different and important functions in the brain and body. For example, serotonin is involved in mood, social behavior, feeding, sleeping, learning and memory, cognition, and emotion regulation. Serotonin has classically been known for its role in mood disorders, including depression and anxiety (for reviews see Belsky & Pluess, 2009; Caspi, Hariri, Holmes, Uher, & Moffitt, 2010; Karg, Burmeister, Shedden & Sen, 2011).

Specifically, the initial investigation of the moderating effect of the 5-HTTLPR cites evidence for 5-HTT moderating reactivity to life stress, thereby bringing about depressive symptoms (Caspi et al., 2003). Since Caspi et al.'s (2003) initial work investigating the relationship between serotonin, stress, and depression, meta-analytic evidence has emerged both in support (Karg et al., 2011; Sharpley, Palanisamy, Glyde, Dillingham, & Agnew, 2014; Bleys, Luyten, Soenens, & Claes, 2018) and in opposition (Munafò, Brown, & Hariri, 2008; Risch et al., 2009) to the hypothesis that the 5-HTTLPR s-allele moderates the relationship between stress and depression. For example, a meta-analysis showed that 5-HTTLPR moderates stress and depression, but only in response to specific types of stress (i.e., childhood maltreatment), specific medical conditions, and not all stressful life events (Karg et al., 2011). Variation in the results from the meta-analysis may be due to the investigation of different interactions, different types of stressors studied, or the use of different statistical modeling (Sharpley et al., 2017). SPS may explain some of the mixed findings.

Some literature suggests that serotonin may be an important marker of SPS. The most direct evidence that the serotonin transporter s-allele of the 5-HTTLPR gene is linked with SPS comes from research with 169 individuals (aged 18–88 years) derived from the Cimbi database (Licht, Mortensen, & Knudsen, 2011; Knudsen et al., 2016). Results specifically showed that higher levels of SPS, measured with the HSP scale (Aron & Aron, 1997), were associated with the s/s genotype. Additionally, a partial regression showed an association between the ease of excitation facet of SPS and the 5-HTTLPR s-allele (Licht et al., 2011).

Further, there is evidence that the 5-HTTLPR mediates differential susceptibility, a hallmark also seen in SPS. Pluess, Belsky, Way, and Taylor (2010) focused on neuroticism, a trait that may mediate the potential association between the 5-HTTLPR s-allele and depression. They found that for those participants homozygous for the short allele, more negative life events were associated with greater neuroticism, whereas more positive life events were associated with less neuroticism. No such association emerged in the case of those homozygous for the long allele. Studies on child development have also demonstrated that the 5-HTTLPR s-allele expresses itself according to the differential susceptibility theories of gene × environment interactions. For instance, adverse childhood experiences increased externalizing and internalizing problems, and a positive childhood decreased these problems (Belsky et al., 2009). Similar evidence comes from two metaanalyses in a review by van Ijzendoorn, Belsky, and Bakermans-Kranenburg (2012). The metaanalyses analyzed the 5-HTTLPR-moderated association between positive environments and developmental outcomes. The results corroborated that carriers of the s-allele of the 5-HTTLPR were more open to adverse as well as positive environments, but only in Caucasian samples. In the total set of studies, including research with mixed ethnicities, carriers with at least one s-allele were significantly more vulnerable to negative environments than l/l allele carriers. In the Caucasian samples, however, carriers with at least one s-allele also profited more from positive environmental input than long/long carriers. The associations between environment and developmental outcomes were absent for l/l carriers. The metaanalytic findings thus support the hypothesis that in Caucasian samples 5-HTTLPR is a genetic marker of differential susceptibility.

The 5-HTTLPR gene is also involved in aspects of social cognition (see review by Canli & Lesch, 2007; Kiser, Steemers, Branchi, & Homberg, 2012) consistent with models of SPS. In humans, two studies reported a significant negative correlation between the 5-HTTLPR s-allele and the personality trait "agreeableness" (Greenberg et al., 2000; Lesch et al., 1996). Social traits, such as agreeableness, may manifest from 5-HTT-linked anxiety, which regulates an individual's responses to actual or threatened social exclusion. In another study where romantic relationships were examined, findings showed that s-allele carriers were more affected by their partner's positive or anxious mood after a marital discussion (Schoebi, Way, Karney, & Bradbury, 2012). Other advantages of having the s/s variant include superior cognitive performance on perceptual tasks, longer reflection before making difficult choices, and better performance on a delayed pattern recognition task (for a review, see Homberg & Lesch, 2011). For instance, 5-HTTLPR s-allele carriers exhibited more risk aversion when there was a low probability of winning, but greater risk-seeking when there was a high probability of winning (Jedema et al., 2010; Roiser, Rogers, Cook, & Sahakian, 2006). Superior cognitive performance in association with 5-HTTLPR has also been evidenced in creativity (Volf, Kulikov, Bortsov, & Popova, 2009). For example, 62 Caucasian subjects with s/s and l/s genotypes demonstrated both higher verbal and figural creativity scores in comparison with the l/l genotype carriers (Volf et al., 2009). This study was the first to report a significant association between the 5-HTTLPR polymorphism and creative achievements.

In sum, the 5-HTTLPR gene effects observed across studies examining its relationship to both mood disorders and positive aspects that contribute to life flourishing—creativity, superior cognitive performance, and enhanced decision making—are in line with SPS theory, differential susceptibility theory, and biological sensitivity to context theories. Collectively, studies across species and developmental stages, showing effects in both children and adults, suggest that 5-HTTLPR s-allele carriers function better psychologically and physically in positive environments than those with the l/l genotype. Conversely, when exposed to adverse environments, individuals with the s-allele are at high-risk for a variety of negative health outcomes, such as anxiety or depression. Although other genes are likely to underlie SPS as discussed later, 5-HTTLPR is a strong candidate gene exemplifying many of the characteristics of high SPS.

2.3 Dopamine

Evidence suggests that dopamine is another major neurotransmitter implicated in modulating SPS. Dopamine is classically known for its role in reward, response to novel stimuli, and motivation (Berke, 2018). With regard to SPS, at least one twin study investigated 16 genes in four different dopamine subsystems among a sample of 478 Chinese university students (Chen et al., 2011). Participants also completed the HSP scale (Aron & Aron, 1997), the parental warmth and acceptance scale (PWAS; Greenberger, Chen, Tally, & Dong, 2000), and the stressful life events scale (Compas, 1987; Greenberger et al., 2000; Wills, Vaccaro, & McNamara, 1992). Results showed that 10 polymorphisms on 7 different genes related to the dopamine system were associated with SPS (Chen et al., 2011). Broadly speaking, the dopamine genes identified in relation to SPS encode for proteins fulfilling various roles in the synthesis of dopamine and epinephrine. However, the exact role and function of these genes in the etiology of SPS remains to be investigated in future research.

Another study directly implicating dopamine in SPS (measured with the HSP Scale) examined 298 healthy Chinese college students' resting-state brain activity (measured with functional MRI) during resting state and genotyped for 98 loci within the dopamine system (Chen et al., 2015). The researchers extracted a "gene score" that summarized the genetic variations representing the 10 loci that were significantly linked to SPS and then used path analysis to search for brain regions whose resting-state data could account for the gene-behavior association. Results showed that the effect of dopamine-related genes was linked to reduced homogeneity of regional spontaneous activity in the precuneus. Regional homogeneity was calculated because its individual differences have been associated with genetic variations. In sum, these results indicate that higher scores indicating more dopamine-related genes were associated with weaker homogeneity of regional spontaneous activity in the precuneus, which in turn led to higher sensitivity.

Research on child development also suggests that dopamine is involved in SPS-related outcomes, such as social adjustment. For example, the 7-repeat allele of the DRD4 (but not 5-HTTLPR) polymorphism moderated the effect of the child-care quality on externalizing problems among 508 Caucasian children at 54 months, and on social skills among children in kindergarten and first grade, but not thereafter (Pluess & Belsky, 2013). The authors investigated whether the effects of day care were genetically moderated by DRD4 and 5-HTTLPR, both genes identified in association with SPS (Chen et al., 2011; Licht et al., 2011). The study reported on social adjustment as measured by repeated teacher reports of externalizing behavior problems (e.g., "hits others," "disobedient at school," "argues a lot") and on social skills (e.g., "makes friends easily," "controls temper when arguing with other children," "asks permission before using someone else's property"). Results showed that the externalizing behaviors proved more consistent with diathesis-stress than differential-susceptibility models, whereas the reverse was true of the social-skills' results (Caspi et al., 2010). As another example, Propper, Willoughby, Halpern, Carbone, and Cox (2007) tested gene × environment interaction effects for both intrusive and warm/responsive maternal behavior. Focusing on externalizing behavior, the association with negative–intrusive parenting was associated with externalizing behaviors, with a slightly larger effect size for children with the DRD4 7-repeat allele. For children with the DRD4 7 VTR (repeat) gene, warm-responsive parenting was associated with less externalizing behavior, whereas for children without the DRD4 7-repeat gene, warm-responsive parenting was associated with more externalizing behavior.

Meta-analytic findings also provide evidence of dopamine's role in differential susceptibility among children. For example, in a meta-analysis examining dopamine gene-environment studies of children up to 10 years old, researchers investigated associations between negative and positive rearing environments and developmental outcomes (Bakermans-Kranenburg & van Ijzendoorn, 2011). Results showed that children with "risky" dopamine (dopamine receptor D2, DRD4, dopamine transporter) genes did worse in various developmental outcomes in negative environments than the comparisons without the "genetic risk." However, children with this particular variant also profited most from positive environments, showing beneficial developmental outcomes.

2.4 Other genes

Other genes that could possibly be associated with SPS due to their association with differential response to early-life stress, parenting quality, emotional stimuli, perceptual processing and mental health interventions include catechol-*O*-methyltransferase (COMT) (Drury et al., 2010), alpha-2B adrenergic receptor (Todd et al., 2015), serotonin receptor 2A (HTRA2) (Jokela et al., 2007), FK506 binding protein 5 (Binder et al., 2008), nerve growth factor (NGF) and brain-derived neurotrophic factor (BDNF) (Lester et al., 2012), mineralocorticoid receptor (Bogdan, Williamson, & Hariri, 2012), μ-opioid receptor (OPRM1) (Troisi et al., 2011), and the glucocorticoid receptor NR3C1 (Albert et al., 2015) genes.

In addition to the 5-HTTLPR variant, other serotonergic genes may be related to SPS. For example, one study found that individuals ages 3–18 carrying the T/T or T/C genotype of the serotonin receptor 2A gene were responsive to the protective aspects of nurturing mothering

(Jokela et al., 2007). Specifically, in the presence of high maternal nurturance, subjects expressed low levels of depressive symptoms, while this was not true with the carriers of the C/C genotype. These findings suggest that further research into the association between serotonin receptor 2A and SPS could be useful.

With respect to BDNF, one study showed a significant protective effect of the met allele on depressive symptoms in children who had been exposed to early severe psychosocial deprivation, measured as institutionalization prior to 31 months of age (Drury et al., 2012). Measured at 54 months of age, depressive symptoms were lower for those with the met/met allele of the BDNF gene. There also was a significant protective effect of the met allele on depressive symptoms within a group who continued in institutionalized care. However, there was no relationship seen with a group assigned to foster care (Drury et al., 2010). These findings suggest that BDNF may be important in explaining resilience, and may also prove to be a candidate marker of SPS.

Another potential candidate gene for SPS, ADRA2b, regulates norepinephrine availability, increases sensitivity to low-level perceptual features, and enhances perceptual vividness (Todd et al., 2015). Although ADRA2b polymorphisms have not been directly investigated in relation to SPS, the association of both the ADRA2b polymorphism and SPS with perceptual sensitivity (Jagiellowicz et al., 2011; Todd et al., 2015) suggests that the perceptual acuity characteristic of SPS may be moderated by ADRA2b.

Although the aforementioned candidate genes may prove to be fruitful avenues of further study, genetic studies have increasingly used a GWAS (genome-wide association study) approach for reasons outlined later. It has been well established that most single genes have very small effect sizes (for review, see Fox & Beevers, 2016). Summing the "risk" or "sensitivity" yielded by several genes provides a quantitative index with a potentially larger effect size than single candidate genes (Fox & Beevers, 2016). Thus, some studies have created a polygenic score of environmental sensitivity based on whole-genome findings and then used the additive score for analyses (Keers et al., 2016; Keers & Pluess, 2017).

Using a GWAS approach, researchers calculated a polygenic score of environmental sensitivity (composed of BDNF, COMT, DRD2, HTRA2, and FKBP5), and measured childhood and adult environmental quality in association with psychological distress in adults (Keers & Pluess, 2017). Results showed evidence for both environment × environment and gene × environment effects on psychological distress. Specifically, genetically sensitive individuals were more vulnerable to adversity as adults if they had experienced a poor childhood environment. However, as adults they were significantly less vulnerable to adversity if their childhood environment had been positive. These finding are in line with the theories of differential susceptibility and provide robust support for gene × environment interactions in the development of psychological issues.

A similar study tested whether a polygenic score of environmental sensitivity predicted response to cognitive behavioral therapy (CBT) in children with anxiety disorders (Keers et al., 2016). Specifically, this study examined the effects of parenting on anxiety disorders in children; and children's response to individual, group, and brief parent-led CBT. Results showed that environmental sensitivity (measured with the polygenic score) significantly moderated the effects of parenting on anxiety disorders and treatment response. Individuals with a high polygenic score (vs a low polygenic score) responded significantly better to individual CBT, than to group CBT or brief parent-led CBT. Thus, this study suggests that individuals with the greatest environmental sensitivity may be most likely to develop anxiety disorders in adverse environments but also benefit more from the most intensive types of treatment.

In sum, 5-HTT in the serotonergic system and some dopamine genes have been identified for their effects on the etiology of high SPS. While these studies represent some significant steps toward understanding the genetic basis of SPS, clearly more research is needed. For example, the polymorphisms identified in the current studies need to be replicated in future studies and it is also pertinent to examine whether other genes also contribute to high sensitivity. As noted, GWAS may be a helpful tool to detect genetic variants (SNPs) hypothesis-free. However, one challenge for these genome wide studies is the large number of participants that is needed to detect genetic variances reliably. As noted, it will be important for future studies to remain open to investigating other systems as it is becoming clear that there may be other genetic variants that affect the expression of SPS.

3 The neural basis of SPS

Other research examining the biological basis of SPS in humans has mostly used functional magnetic resonance imaging (fMRI), which detects changes in brain activity by measuring blood oxygenation level signals (BOLD signal), to examine neural responsivity to various stimuli. Studies with rodents have also started to contribute to our understanding of the neural pathways that modulate SPS.

3.1 Neurobiology of SPS: Evidence from human studies

In recent decades, researchers have started to investigate the neural underpinnings of SPS in response to a variety of stimuli from images of landscapes, to general emotionally evocative images; and even using a paradigm to elicit empathy in response to a spouse or a stranger. This body of work, using fMRI, has revealed that SPS is associated with greater activation of neural regions involved in higher-order visual processing, attention, awareness, integration of information, memory, and readiness to act (for a review, see Acevedo, Aron, Pospos, & Jessen, 2018). We discuss these studies below.

In the first study examining the neural basis of SPS, researchers compared differences in brain activity in response to context-dependent and context-independent tasks as a function of SPS across cultures (Aron et al., 2010). The study compared brain activity across Asian participants, generally mastering context-dependent tasks more readily; and American participants, generally more at ease with context-independent tasks. Among participants scoring high on the HSP scale, the differences in brain activity between context-dependent and context-independent tasks were subtle, regardless of the individual's cultural background. This finding indicates less biased processing for high SPS suggesting that task performance in HSPs is less dependent on cultural norms, while task performance in less sensitive individuals is more dependent on cultural norms (Aron et al., 2010). These results may also suggest an increased openness to all environmental stimuli among high SPS individuals, such that they process all stimuli deeply, independent of the social and cultural norms they grew up in.

In another study examining the neural basis of SPS, 18 participants were instructed to engage in a visual discrimination task and to notice differences between pictures showing major and minor visual changes (Jagiellowicz et al., 2011). A minor change was exemplified by the addition of half a haystack to the front of a row of haystacks, whereas a major change was exemplified by the addition of a fence post into a sequence of fence posts. When viewing the images while in the fMRI scanner, participants pressed one button to indicate if the image was the same as the preceding image, and another button to indicate that the images were different. The results showed that areas involved in attention and higher-order visual processing were increasingly activated in association with higher HSP scores. These areas included the medial and posterior parietal regions, the temporal region, and the left occipitotemporal region. When it was necessary to successfully complete a task, highly sensitive individuals tended to take more time to make fine distinctions between images, with HSP scores predicting the time taken to complete the task. Perhaps this is due to their paying more attention to and making subtle distinctions across stimuli.

In yet another fMRI study examining the neural basis of SPS, a cardinal feature of SPS was investigated: heightened responsiveness to social/emotional stimuli. In the first fMRI study to examine SPS and neural responsivity to others' emotional displays, Acevedo et al. (2014) scanned a group of individuals twice (about one year apart) while viewing happy, sad, or neutral face images of a spouse or a stranger. Prior to each face image, participants were shown a short sentence describing the event, such as "Your partner is feeling very happy because something wonderful has happened to them," or "Your partner is sad and suffering because something terrible has happened to them." A corresponding text description was shown prior to each face image because context plays an important role in emotion processing (Gross & John, 2003), and thus by reducing cognitive ambiguity, emotion-specific effects could be reliably measured (e.g., McRae, Misra, Prasad, Pereira, & Gross, 2012). The overall task was adapted from research on maternal responsiveness (Strathearn, Li, Fonagy, & Montague, 2008), and it was utilized to elicit empathy in the context of romantic relationships in the study by Acevedo et al. (2014).

Results of this study showed that SPS was associated with significant activation of brain regions involved in attention, emotion, cognition, and preparation for action. These activations were shown across targets (partners and strangers), emotions (happy and sad), and time (as individuals were scanned twice, about a year apart). Other significant activations were found in areas associated with awareness, sensory integration, empathy, and intuition: namely, the insula and inferior frontal gyrus (IFG) (Acevedo et al., 2014). Perceiving and integrating internal sensory stimuli with current emotional states, the insula is associated with emotions and feelings related to bodily experiences (Craig, 2009). Insula activation is also associated with greater awareness of the inner and outer environment and is thought to be associated with consciousness (Craig, 2009). Also, activity in the cingulate cortex (associated with coordinating attention and awareness), and the premotor area (involved in action planning and unconscious behavioral control) showed stronger activity as a function of higher SPS scores.

Collectively, results reported by Acevedo et al. (2014) highlight the neural correlates underlying some of the cardinal features of SPS, such as increased awareness and intuition, which involve perceiving and "feeling" as well as responding to others' emotional needs, depth of cognitive processing, and integration of information from multisensory modalities. Furthermore, consistent with theories of differential susceptibility (Belsky, 2016), participants showed differential patterns of neural activity as a function of SPS in response to partners' (vs strangers') positive (happy) and negative (sad) face images. For example, SPS was positively associated with neural response to partner versus stranger face images in brain regions implicated in self-other processing, such as the angular gyrus (AG). Although this suggests a differential responsivity to a partner versus a stranger, at the same time there was activation in areas related to empathy, imitation, and self-other processing (such as the IFG and AG). Interestingly, these were both activated in response to partner-happy and stranger-happy conditions, but not for the stranger-sad condition. These results suggest a bias toward the processing of positive emotions among highly sensitive persons. Overall, these strategies highlight the positive side of SPS as one may reap benefits from vicariously experiencing and integrating others' positive experiences in the self, thus highlighting one of the advantages of being highly sensitive.

Yet another study investigated the neural correlates of SPS (using the HSP Scale) in response to general positive, negative, and neutral images from the International Affective Picture System—IAPS (Lang, Bradley, & Cuthbert, 1997) among females scoring high and low on the HSP Scale (Acevedo et al., 2017). Participants also completed a neuroticism scale and measures of perceived childhood quality [asking questions about the parents' temperament (nurturing, harsh, etc.) and the home environment (calm, harsh, etc.)]. Results showed that SPS (and its interaction with a positive childhood environment) was significantly correlated with neural activity in areas involved in memory, emotion, hormonal balance, and reflective thinking (the hippocampus, entorhinal area, hypothalamus, and temporal/parietal areas, respectively). Furthermore, results showed that SPS was associated with a stronger reward response (in the ventral tegmental area and nucleus accumbens) to positive stimuli, and this effect was especially amplified for individuals reporting higher-quality childhoods.

Acevedo et al. (2017) also noted that highly sensitive individuals showed significant decreases in reward areas when viewing negative images. However, high SPS individuals reporting a positive childhood did not exhibit this decreased reward response to negative stimuli. Instead, high SPS individuals with positive childhoods showed stronger activation in brain regions implicated in emotion processing and self-regulation (i.e., the amygdala and prefrontal cortex), without diminished reward activity (which was seen for the simple correlation with SPS), in response to negative stimuli. These results highlight the importance of childhood rearing environment and early development, especially among highly sensitive individuals, for emotion and self-regulation in adulthood. The results of this study also show how via the brain, having a positive childhood may have a long-term impact on individuals' automatic response at the neural level. In sum, these results highlight one of the critical mechanisms affected by rearing environment and expression of SPS through adulthood: self-regulation. Through self-regulation, high HSP individuals (with positive childhoods) neural sysems were able to readily modify dampened reward effects in response to negative stimuli. That is, these individuals have learned (integrated) adaptive regulatory mechanisms for responding to stimuli that otherwise may be perceived as threatening or stressful. Again, here we highlight another positive aspect of the trait, but in tandem with a positive childhood-rearing environment.

In summary, fMRI studies of SPS in humans are consistent with theories highlighting depth of processing and responsiveness to others' emotional signals. For example, across all imaging studies, greater SPS was associated with increased activation of brain regions that are involved in reward processing, memory, emotion, empathy/self-other processing, and awareness. These activations reflect fundamental features of SPS, including depth of processing, awareness, attention, emotion, and responsiveness. This body of work has advanced our basic understanding of the neural mechanisms underlying SPS and has also added to greater characterization of this trait. However, brain regions do not function in isolation. Thus, it's important to consider other neural analytic procedures such as brain connectivity examining specific regions and large-scale brain networks.

3.2 Neurobiology of SPS: Evidence from animal studies

Although SPS studies of the human brain have led to marked discoveries, there are several practical and ethical limitations to human research. Laboratory animals allow control over environmental factors and genetic factors, which more readily permit investigation of the causal mechanisms underlying SPS. This is especially important when aiming to investigate how central cellular and molecular changes shape SPS. In laboratory animals, one can even detect fine electrical signals from single cells, giving a much more precise view of how single neurons or brain regions communicate (Bennett & Ringach, 2016; Institute of Medicine and National Research Council, 2012). In the following content, we discuss SPS-related brain function in animal studies examining mechanisms at the molecular level.

One potential animal model for the 5-HTTLPR in humans involves 5-HTT knockout mice and rats (Caspi et al., 2010). These animals show phenotypes in line with SPS. For example, they display increased emotionality (both negative and positive), decreased exploration of open spaces, and increased responding for sucrose reward (Homberg, Schubert, Asan, & Aron, 2016). 5-HTT knockout mice also show deep information processing, or "pause-to-check" behaviors. When exposed to a cue previously paired with shock in the fear conditioning test, they show increased freezing. Freezing is not just passive fear behavior, it is a brake on the motor system allowing the preparation of action. Hence, freezing, like "pause-to-check," is a highly adaptive behavior subserving the processing of information in order to make the best response in the given situation. Further, the animals display increased sensitivity to environmental subtleties. This is expressed by reduced latent inhibition, which indicates that the animals do not show learned ignorance, instead they remain open to environmental stimulation. Last, the animals showed increased anxiety- and depression-like phenotypes when they were exposed to stressful situations.

Beyond the behavioral level, the animals also have been tested for underlying neurobiological mechanisms, which can help to further direct hypotheses regarding the neurobiological mechanisms underlying the changes in brain activation observed in humans. This is critical, as brain regions are highly heterogeneous, consisting of many types of different neurons. Brain activity as measured by MRI in humans reflects a net change in firing rates of the individual neurons, whereby it is also possible that increases and decreases in neuronal activity rule each other out at MRI level. Since animal studies allow both brain-wide analyses using techniques comparable to those used in humans, as well as fine-grained invasive brain analyses, they offer a great medium to understand brain mechanisms in more detail.

One study used a radiotracer to track brain activity in 5-HTT knockout mice. The tracer was infused in vivo, after which the animals were sacrificed to determine by autoradiography brain region activity with high resolution ex vivo (Pang et al., 2011). The study delivered results, which paralleled those found in humans. Comparing 5-HTT knockout mice to wild-type mice exposed to fear-conditioning tests, results showed that altered brain activity was significantly related to the 5-HTT knockout genotype in the frontal association cortex, motor, and (somato)sensory cortices, insula and amygdala (increased activity), and nucleus accumbens (decreased activity). In addition to genotype, an effect of fear conditioning was observed. More specifically, 5-HTT knockout rats displayed more freezing behavior during the fear conditioning test, and this was associated with a neural activity pattern that was grossly the same as for the naïve animals. However, the activity of the nucleus accumbens switched from reduced activity under naïve conditions, to increased activity during fear extinction (Pang et al., 2011). These results seem to be in line with findings in humans, as individuals high in SPS also exhibited a decreased activation of reward regions in response to a negative stimulus.

In relation to the perception of environmental subtleties, the excitation-inhibition balance was investigated in the barrel cortex, the part of the primary somatosensory cortex where whiskers are represented. The behavioral test involved the so-called gap crossing task. In this task, animals had to cross a gap between two platforms in the dark by using their whiskers. The ability to perceive the platform at the other side of the gap at different distances requires fine sensory perception. It was found that the 5-HTT knockout rats needed less whisker touches to decide to cross the gap. They were also faster in crossing the gasp. This suggests greater sensory integration (Miceli et al., 2017). Subsequently, electrophysiology and molecular techniques were used to study the underlying mechanisms in the barrel cortex in terms of the excitation-inhibition balance. Researchers found that the inhibitory control over excitatory neurons in the barrel cortex was reduced, leading to hyperexcitability of the glutamatergic excitatory neurons (Miceli et al., 2017). This could reflect a mechanism contributing increased activity of a brain region, exactly as was found in the study of Pang et al. (2011) discussed earlier using autoradiography. Hence, the activity of brain regions could be increased because the inhibitory brake is working less. While the measurements focused here on the barrel cortex, there is evidence for a reduction in GABAergic inhibitory signaling in the hippocampus and prefrontal cortex as well (Luoni et al., 2013; Guidotti et al., 2012). In line with the increased excitability observed in 5-HTT knockout rats, Nietzer et al. (2011) observed that the number of spines on excitatory pyramidal neurons in the amygdala was increased in 5-HTT knockout mice. This is indicative of increased excitability. Interestingly, whereas stress increased the number of spines in wild-type mice up to the level observed in knockout mice, there was no further increase in the number of spines in the knockout mice, probably due to a ceiling effect. Behaviorally, the 5-HTT knockout but not wild-type animals are affected by stress. Since stress reduces inhibitory control and increases excitation, it can be speculated that if inhibitory control is already low, and excitability is already high at baseline, then the ability to cope with stress by reducing inhibition and increasing excitation is reduced, leading to overstimulation, stress, and eventually increased risk for stress-related disorders. This may provide a basis for the increased stress sensitivity seen in 5-HTT knockout animals and individuals high in SPS. See Homberg et al. (2016) for a further discussion of this hypothesis.

In summary, phenotypes of 5-HTT knockout animals resemble those observed in human 5-HTTLPR s-allele carriers, as well as those seen in humans high in SPS. The animals display brain activity changes that are also seen in high-SPS. Furthermore, increased activity for the somatosensory cortex appears to be related to reduced inhibitory control over excitatory neurons. Hence, reduced inhibitory control, rather than increased neuronal activity per se, may underlie increased activity of some brain regions thus resulting in some of the characteristic expressions of high SPS.

4 Peripheral sensitivities

SPS is most likely not restricted to the central nervous system but may extend to other parts of the body, resulting in alterations in hormonal balance, glucose metabolism, enhanced sensations, etc. There are several theories and pathways that may mediate peripheral sensitivity in SPS, which we outline now.

4.1 Hypothalamus-pituitary-adrenal axis

One pathway that may mediate peripheral sensitivities in SPS is the hypothalamus-pituitary (HPA)-axis, which is classically known for its role in stress response. Briefly, the paraventricular nucleus of the hypothalamus contains neuroendocrine neurons that synthesize and secrete corticotropin-releasing hormone (CRH). This peptide stimulates the secretion of adrenocorticotropic hormone (ACTH) from the anterior lobe of the pituitary gland. ACTH in turn acts on the adrenal cortex, which produces glucocorticoid hormones (mainly cortisol in humans and corticosterone in rodents). Glucocorticoids in turn act on several organs to regulate energy and prepare the body for action, and act back on the hypothalamus and pituitary to suppress CRH and ACTH production in a negative feedback cycle.

An animal study that used 5-HTT knockout rats to examine functioning of the HPA-axis (van der Doelen et al., 2014) found that at baseline, the knockout animals showed increased plasma corticosterone levels and increased expression of the ACTH receptor in the adrenal cortex. Additionally, a higher sensitivity to ACTH was found in the adrenal glands. However, no changes were found in other parts of the HPA-axis. These findings suggest involvement of the peripheral part of the HPA-axis in regulating internal processes related to SPS (van der Doelen et al., 2014).

4.2 Glucose metabolism and sensations

Apart from alterations in the HPA-axis found in 5-HTT knockout rats, humans with high SPS have been reported to be very sensitive to spikes and dips in blood sugar, and also react strongly to the feeling of hunger (Aron et al., 2010). This could be explained by alterations in the neurotensin system (including neurolysin) as shown in the polymorphisms found by Chen et al. (2011). For example, neurolysin is implicated in glucose metabolism (Cavalcanti et al., 2013). As such, it may be implicated in the higher frequency of SPS observed in adolescents with type 1 diabetes as compared to controls (Goldberg et al., 2018).

Neurolysin is also implicated in pain control (Feng, Wang, Dong, Wang, & Li, 2015). Thus, it may account for lower pain tolerance reported by individuals with high SPS (Aron et al., 2010). Both SPS and the chronic pain condition of fibromyalgia (a disorder characterized by widespread musculoskeletal pain accompanied by fatigue, sleep issues, memory loss, and depressed mood) are associated with the s-allele of the 5-HTTLPR (Offenbaecher et al., 1999), suggesting that alterations in serotonergic function may also contribute to increased sensitivity to pain and high SPS.

Additionally, the periaqueductal gray (PAG)—a brain region that is involved in pain-processing, regulation of anxiety, and autonomic responses (Bittar, Nandi, Carter, & Aziz, 2005) may also be a key neural correlate of pain processing in fibromyalgia (Bennett, 1999) and sensitivity to stimuli among highly sensitive persons. 5-HTT knockout rats show altered activity of GABAergic neurons that project from the amygdala to the PAG, raising a potential mechanism through which 5-HTT gene variance may influence the perception of threatening stimuli (Schipper et al., 2019).

5 Summary and future perspectives

SPS is a complex trait influenced by a great variety of factors, namely, environmental and genetic etiological factors. Research has started to identify the neural and genetic factors underlying SPS. To date, the serotonin transporter gene (5-HTTLPR) and a number of dopamine genes have been identified as reliable markers of SPS. However, further research is required to identify other genetic variants and to confirm present findings. GWAS (genome wide association study) is one approach that may be useful to identify other genetic variants and polymorphisms that underlie the etiology of SPS. The value of GWAS, as noted, is that gene variance across the entire genome is being assessed in relation to the trait under investigation.

This provides a hypothesis-free approach and allows the identification of genes associated with a given trait or disorder that may not be among the candidate genes. Another promising approach may be to use a polygenic score, developed by first doing a GWAS, on a discovery sample to identify potential candidate genes, then weighting the candidate genes to calculate a genetic index (i.e., a polygenic sensitivity score), which is then applied to each individual in a validation sample.

Research on 5-HTT knockout animals that model the human 5-HTTLPR short allele, show increased integration of sensory information, which is related to reduced inhibitory control over excitatory neurons in the somatosensory cortex in mice. This process might explain the alertness and responsivity to subtle stimuli observed in high SPS. Thus, research with 5-HTT knockout mice has shown some of the possible molecular mechanisms mediating SPS.

Human brain imaging research has also provided strong biological evidence for features observed in highly sensitive persons (HSPs), such as enhanced awareness, empathy, and depth of processing. More specifically, fMRI studies with humans exposed to a variety of stimuli (from scenes, emotional faces, emotional objects, and perceptual tasks) have all identified the parietal lobe in SPS processing (Acevedo et al., 2018). Also, a review of the brain regions implicated in SPS processing in comparison with seemingly related clinical disorders (including autism, schizophrenia, and posttraumatic stress disorder) reported that SPS uniquely showed activations in neural structures associated with) physiological homeostasis and pain control (hypothalamus and PAG), self-other processing and empathy (inferior frontal gyrus and insula), awareness and reflective thinking (TPJ), and self-control (prefrontal cortex). This body of work has also provided support for differential susceptibility as HSPs showed a strong reward response (in the ventral tegmental area) to positive stimuli (such as a partner smiling).

As SPS in humans is characterized by increased awareness of environmental subtleties, emotionality, and in-depth information processing, it would be useful to develop animal models for SPS that align with these behavioral features (see Section 2). Such a "phenotypic" animal models may overcome the small effect size of single genes and increase the translational value of animal-derived data. After establishing new animal models, neural differences can be compared between high SPS and low SPS populations focusing on cellular alterations underlying SPS. In animals, it is possible to manipulate specific brain circuits and trace neural activation of single neurons establishing precise functional networks. Thus, in animals, functional anatomical mapping can be linked to specific SPS phenotypes to understand whether overlapping or different patterns of neural circuitry differentially shape the different SPS facets.

In conclusion, the relatively few but growing studies on the neural and genetic basis of SPS provide evidence of the neurobiological mechanisms underlying SPS, which give rise to its core features like enhanced awareness, responsivity, empathy, emotional reactivity, and deep information processing. This body of work also highlights the impact of childhood environment on long-term outcomes such as depression and anxiety, and on the positive side enhanced cognitive performance and better self-regulation. However, as noted, future work remains to be done to refine and expand our scientific understanding of both the etiology and the biological mechanisms rendering some individuals more sensitive to environmental influences than others.

References

Acevedo, B., Aron, E., Pospos, S., & Jessen, D. (2018). The functional highly sensitive brain: A review of the brain circuits underlying sensory processing sensitivity and seemingly related disorders. *Philosophical Transactions of the Royal Society of London. Series B, Biological Sciences*, *373*(1744), 5. https://doi.org/10.1098/rstb.2017.0161.

Acevedo, B. P., Aron, E. N., Aron, A., Sangster, M. D., Collins, N., & Brown, L. L. (2014). The highly sensitive brain: An fMRI study of sensory processing sensitivity and response to others' emotions. *Brain and Behavior*, *4*(4), 580–594. https://doi.org/10.1002/brb3.242.

Acevedo, B. P., Jagiellowicz, J., Aron, E., Marhenke, R., & Aron, A. (2017). Sensory processing sensitivity and childhood quality's effects on neural responses to emotional stimuli. *Clinical Neuropsychiatry*, *14*(6), 359–373.

Albert, D., Belsky, D. W., Crowley, D. M., Bates, J. E., Pettit, G. S., Lansford, J. E., Dick, D., & Dodge, K. A. (2015). Developmental mediation of genetic variation in response to the Fast Track prevention program. *Development and Psychopathology*, *27*(1), 81–95.

Aron, A., Ketay, S., Hedden, T., Aron, E. N., Rose Markus, H., & Gabrieli, J. D. (2010). Temperament trait of sensory processing sensitivity moderates cultural differences in neural response. *Social Cognitive and Affective Neuroscience*, *5*(2–3), 219–226. https://doi.org/10.1093/scan/nsq028.

Aron, E. N., & Aron, A. (1997). Sensory-processing sensitivity and its relation to introversion and emotionality. *Journal of Personality and Social Psychology*, *73*(2), 345–368.

Assary, E., Zavos, H. M., Krapohl, E., & Pluess, M. (2019). Heritability of environmental sensitivity and genetic overlap with the big-five personality traits (under review).

Bakermans-Kranenburg, M. J., & van Ijzendoorn, M. H. (2011). Differential susceptibility to rearing environment depending on dopamine-related genes: New evidence and a meta-analysis. *Development and Psychopathology*, *23*(1), 39–52. https://doi.org/10.1017/S0954579410000635.

Belsky, J. (1997). Variation in susceptibility to environmental influence: An evolutionary argument. *Psychological Inquiry*, *8*(3), 182–186. https://doi.org/10.1207/s15327965pli0803_3.

Belsky, J. (2016). The differential susceptibility hypothesis: Sensitivity to the environment for better and for worse. *JAMA Pediatrics*, *170*(4), 321–322. https://doi.org/10.1001/jamapediatrics.2015.4263.

Belsky, J., Hsieh, K. H., & Crnic, K. (1998). Mothering, fathering, and infant negativity as antecedents of boys' externalizing problems and inhibition at age 3 years: Differential susceptibility to rearing experience? *Development and Psychopathology*, *10*(2), 301–319. https://doi.org/10.1017/s095457949800162x.

Belsky, J., Jonassaint, C., Pluess, M., Stanton, M., Brummett, B., & Williams, R. (2009). Vulnerability genes or plasticity genes? *Molecular Psychiatry*, *14*(8), 746–754. https://doi.org/10.1038/mp.2009.44.

Belsky, J., & Pluess, M. (2009). Beyond diathesis stress: Differential susceptibility to environmental influences. *Psychological Bulletin*, *135*(6), 885–908. https://doi.org/10.1037/a0017376.

Bennett, A. J., & Ringach, D. L. (2016). Animal research in neuroscience: A duty to engage. *Neuron*, *92*(3), 653–657. https://doi.org/10.1016/j.neuron.2016.10.034.

Bennett, R. M. (1999). Emerging concepts in the neurobiology of chronic pain: Evidence of abnormal sensory processing in fibromyalgia. *Mayo Clinic Proceedings*, *74*(4), 385–398. https://doi.org/10.4065/74.4.385.

Berke, J. D. (2018). What does dopamine mean? *Nature Neuroscience*, *21*(6), 787–793. https://doi.org/10.1038/s41593-018-0152-y.

Binder, E. B., Bradley, R. G., Liu, W., Epstein, M. P., Deveau, T. C., Mercer, K. B., Tang, Y., Gillespie, C. F., Heim, C. M., Nemeroff, C. B., Schwartz, A. C., Cubells, J. F., & Ressler, K. J. (2008). Association of FKBP5 polymorphisms and childhood abuse with risk of posttraumatic stress disorder symptoms in adults. *JAMA*, *299*(11), 1291–1305.

Bittar, R. G., Nandi, D., Carter, H., & Aziz, T. Z. (2005). Somatotopic organization of the human periventricular gray matter. *Journal of Clinical Neuroscience*, *12*(3), 240–241. https://doi.org/10.1016/j.jocn.2004.05.012.

Bleys, D., Luyten, P., Soenens, B., & Claes, S. (2018). Gene-environment interactions between stress and 5-HTTLPR in depression: A meta-analytic update. *Journal of Affective Disorders*, *226*, 339–345. https://doi.org/10.1016/j.jad.2017.09.050.

Bogdan, R., Williamson, D. E., & Hariri, A. R. (2012). Mineralocorticoid receptor Iso/Val (rs5522) genotype moderates the association between previous childhood emotional neglect and amygdala reactivity. *The American Journal of Psychiatry*, *169*(5), 515–522.

Bridges, D., & Schendan, H. E. (2018a). Sensitive individuals are more creative. *Personality and Individual Differences*, https://doi.org/10.1016/j.paid.2018.09.015.

Bridges, D., & Schendan, H. E. (2018b). The sensitive, open creator. *Personality and Individual Differences*, *142*, 179–185.

Canli, T., & Lesch, K. P. (2007). Long story short: The serotonin transporter in emotion regulation and social cognition. *Nature Neuroscience*, *10*(9), 1103–1109.

Carter, C. S., Grippo, A. J., Pournajafi-Nazarloo, H., Ruscio, M. G., & Porges, S. W. (2008). Oxytocin, vasopressin and sociality. *Progress in Brain Research*, *170*, 331–336. https://doi.org/10.1016/S0079-6123(08)00427-5.

Caspi, A., Hariri, A. R., Holmes, A., Uher, R., & Moffitt, T. E. (2010). Genetic sensitivity to the environment: The case of the serotonin transporter gene and its implications for studying complex diseases and traits. *The American Journal of Psychiatry*, *2167*(5), 509–527. https://doi.org/10.1176/appi.ajp.2010.09101452.

Caspi, A., Sugden, K., Moffitt, T. E., Taylor, A., Craig, I. W., Harrington, H., … Poulton, R. (2003). Influence of life stress on depression: Moderation by a polymorphism in the 5-HTT gene. *Science*, *301*(5631), 386–389. https://doi.org/10.1126/science.1083968.

Cavalcanti, D., Barros, C. C., Castro, L. M., Neto, J. C. R., Shadock, I., Bader, M., & Ferro, E. S. (2013). Glucose and intracellular peptides metabolism alteration in neurolysin knockout mice. *FEBS Journal*, *280*, 496.

Chen, C. H., Chen, C. S., Moyzis, R., Stern, H., He, Q. H., Li, H., … Dong, Q. (2011). Contributions of dopamine-related genes and environmental factors to highly sensitive personality: A multi-step neuronal system-level approach. *PLoS One*, *6*(7), 9. https://doi.org/10.1371/journal.pone.0021636.

Chen, C., Xiu, D., Chen, C., Moyzis, R., Xia, M., He, Y., Xue, G., Li, J., He, Q., Lei, X., Wang, Y., Liu, B., Chen, W., Zhu, B., , … Dong, Q. (2015). Regional homogeneity of resting-state brain activity suppresses the effect of dopamine-related genes on sensory processing sensitivity. *PLoS One*, *10*(8), e0133143.

Compas, B. E. (1987). Stress and life events during childhood and adolescence. *Clinical Psychology Review*, 7(3), 275–302. https://doi.org/10.1016/0272-7358(87)90037-7.

Coppens, C. M., de Boer, S. F., & Koolhaas, J. M. (2010). Coping styles and behavioural flexibility: Towards underlying mechanisms. *Philosophical Transactions of the Royal Society of London Series B, Biological Sciences*, *365*(1560), 4021–4028. https://doi.org/10.1098/rstb.2010.0217.

Craig, A. D. (2009). How do you feel—Now? The anterior insula and human awareness. *Nature Reviews Neuroscience*, *10*(1), 59–70. https://doi.org/10.1038/nrn2555.

Drury, S. S., Gleason, M. M., Theall, K. P., Smyke, A. T., Nelson, C. A., Fox, N. A., & Zeanah, C. H. (2012). Genetic sensitivity to the caregiving context: The influence of 5httlpr and BDNF val66met on indiscriminate social behavior. *Physiology & Behavior*, *106*(5), 728–735. https://doi.org/10.1016/j.physbeh.2011.11.014.

Drury, S. S., Theall, K. P., Smyke, A. T., Keats, B. J., Egger, H. L., Nelson, C. A., Fox, N. A., Marshall, P. J., & Zeanah, C. H. (2010). Modification of depression by COMT val158met polymorphism in children exposed to early severe psychosocial deprivation. *Child Abuse & Neglect*, *34*(6), 387–395. https://doi.org/10.1016/j.chiabu.2009.09.021.

Feng, Y. P., Wang, J., Dong, Y. L., Wang, Y. Y., & Li, Y. Q. (2015). The roles of neurotensin and its analogues in pain. *Current Pharmaceutical Design*, *21*(7), 840–848.

Forssman, L., Peltola, M. J., Yrttiaho, S., Puura, K., Mononen, N., Lehtimäki, T., & Leppänen, J. M. (2014). Regulatory variant of the TPH2 gene and early life stress are associated with heightened attention to social signals of fear in infants. *Journal of Child Psychology and Psychiatry*, *55*(7), 793–801. https://doi.org/10.1111/jcpp.12181.

Fox, E., & Beevers, C. G. (2016). Differential sensitivity to the environment: Contribution of cognitive biases and genes to psychological wellbeing. *Molecular Psychiatry*, *21*(12), 1657–1662. https://doi.org/10.1038/mp.2016.114.

Goldberg, A., Ebraheem, Z., Freiberg, C., Ferarro, R., Chai, S., & Gottfried, O. D. (2018). Sweet and sensitive: Sensory processing sensitivity and type 1 diabetes. *Journal of Pediatric Nursing*, *38*, E35–E38. https://doi.org/10.1016/j.pedn.2017.10.015.

Greenberger, E., Chen, C. S., Tally, S. R., & Dong, Q. (2000). Family, peer, and individual correlates of depressive symptomatology among US and Chinese adolescents. *Journal of Consulting and Clinical Psychology*, *68*(2), 209–219. https://doi.org/10.1037//0022-006x.68.2.209.

Greenberg, B. D., Li, Q., Lucas, F. R., Hu, S., Sirota, L. A., Benjamin, J., Lesch, K. P., Hamer, D., & Murphy, D. L. (2000). Association between the serotonin transporter promoter polymorphism and personality traits in a primarily female population sample. *American Journal of Medical Genetics*, *96*(2), 202–216.

Gross, J. J., & John, O. P. (2003). Individual differences in two emotion regulation processes: Implications for affect, relationships, and well-being. *Journal of Personality and Social Psychology*, *85*(2), 348–362.

Guidotti, G., Calabrese, F., Auletta, F., Olivier, J., Racagni, G., Homberg, J., & Riva, M. A. (2012). Developmental influence of the serotonin transporter on the expression of npas4 and GABAergic markers: Modulation by antidepressant treatment. *Neuropsychopharmacology*, *37*(3), 746–758. https://doi.org/10.1038/npp.2011.252.

Hartman, S., Freeman, S. M., Bales, K. L., & Belsky, J. (2018). Prenatal stress as a risk- and an opportunity-factor. *Psychological Science*, *29*(4), 572–580. https://doi.org/10.1177/0956797617739983.

Hessing, M. J., Hagelsø, A. M., Schouten, W. G., Wiepkema, P. R., & van Beek, J. A. (1994). Individual behavioral and physiological strategies in pigs. *Physiology & Behavior*, *55*(1), 39–46.

Homberg, J. R., & Lesch, K. P. (2011). Looking on the bright side of serotonin transporter gene variation. *Biological Psychiatry*, *69*(6), 513–519. https://doi.org/10.1016/j.biopsych.2010.09.024.

Homberg, J. R., Schubert, D., Asan, E., & Aron, E. N. (2016). Sensory processing sensitivity and serotonin gene variance: Insights into mechanisms shaping environmental sensitivity. *Neuroscience and Biobehavioral Reviews*, *71*, 472–483. https://doi.org/10.1016/j.neubiorev.2016.09.029.

Institute of Medicine and National Research Council. (2012). *International animal research regulations: Impact on neuroscience research: Workshop summary*. Washington, DC: The National Academies Press.

Jagiellowicz, J., Xu, X. M., Aron, A., Aron, E., Cao, G. K., Feng, T. Y., & Weng, X. C. (2011). The trait of sensory processing sensitivity and neural responses to changes in visual scenes. *Social Cognitive and Affective Neuroscience*, *6*(1), 38–47. https://doi.org/10.1093/scan/nsq001.

Jedema, H. P., Gianaros, P. J., Greer, P. J., Kerr, D. D., Liu, S., Higley, J. D., … Bradberry, C. W. (2010). Cognitive impact of genetic variation of the serotonin transporter in primates is associated with differences in brain morphology rather than serotonin neurotransmission. *Molecular Psychiatry*, *15*(5), 512–522. 446 https://doi.org/10.1038/mp.2009.90.

Jokela, M., Keltikangas-Järvinen, L., Kivimäki, M., Puttonen, S., Elovainio, M., Rontu, R., & Lehtimäki, T. (2007). Serotonin receptor 2A gene and the influence of childhood maternal nurturance on adulthood depressive symptoms. *Archives of General Psychiatry*, *64*(3), 356–360.

Karg, K., Burmeister, M., Shedden, K., & Sen, S. (2011). The serotonin transporter promoter variant (5-HTTLPR), stress, and depression meta-analysis revisited: Evidence of genetic moderation. *Archives of General Psychiatry*, *68*(5), 444–454. https://doi.org/10.1001/archgenpsychiatry.2010.189.

Keers, R., Coleman, J. R., Lester, K. J., Roberts, S., Breen, G., Thastum, M., Bögels, S., Schneider, S., Heiervang, E., Meiser-Stedman, R., Nauta, M., Creswell, C., Thirlwall, K., Rapee, R. M., Hudson, J. L., Lewis, C., Plomin, R., & Eley, T. C. (2016). A genome-wide test of the differential susceptibility hypothesis reveals a genetic predictor of differential response to psychological treatments for child anxiety disorders. *Psychotherapy and Psychosomatics*, *85*(3), 146–158. https://doi.org/10.1159/000444023.

Keers, R., & Pluess, M. (2017). Childhood quality influences genetic sensitivity to environmental influences across adulthood: A life-course gene × environment interaction study. *Development and Psychopathology*, *29*(5), 1921–1933.

Kiser, D., Steemers, B., Branchi, I., & Homberg, J. R. (2012). The reciprocal interaction between serotonin and social behaviour. *Neuroscience and Biobehavioral Reviews*, *36*(2), 786–798. https://doi.org/10.1016/j.neubiorev.2011.12.009.

Knudsen, G. M., Jensen, P. S., Erritzoe, D., Baare, W. F. C., Ettrup, A., Fisher, P. M., … Frokjaer, V. G. (2016). The center for integrated molecular brain imaging (Cimbi) database. *NeuroImage*, *124*, 1213–1219. https://doi.org/10.1016/j.neuroimage.2015.04.025.

Lang, P., Bradley, M., & Cuthbert, B. (1997). *International Affective Picture System (IAPS): Technical manual and affective ratings*. NIMH Center for the Study of Emotion and Attention.

Lesch, K. P., Bengel, D., Heils, A., Sabol, S. Z., Greenberg, B. D., Petri, S., Benjamin, J., Müller, C. R., Hamer, D. H., & Murphy, D. L. (1996). Association of anxiety-related traits with a polymorphism in the serotonin transporter gene regulatory region. *Science*, *274*, 1527–1531.

Lester, K. J., Hudson, J. L., Tropeano, M., Creswell, C., Collier, D. A., Farmer, A., Lyneham, H. J., Rapee, R. M., & Eley, T. C. (2012). Neurotrophic gene polymorphisms and response to psychological therapy. *Translational Psychiatry*, *2*, e108. https://doi.org/10.1038/tp.2012.33.

Licht, C. L., Mortensen, E. L., & Knudsen, G. M. (2011). Association between sensory processing sensitivity and the serotonin transporter polymorphism 5-HTTLPR short/short genotype. *Biological Psychiatry*, *69*(9), 152S–153S.

Luoni, A., Hulsken, S., Cazzaniga, G., Racagni, G., Homberg, J. R., & Riva, M. A. (2013). Behavioural and neuroplastic properties of chronic lurasidone treatment in serotonin transporter knockout rats. *The International Journal of Neuropsychopharmacology*, *16*(6), 1319–1330. https://doi.org/10.1017/S1461145712001332.

Lyons, D. M., Price, E. O., & Moberg, G. P. (1988). Social modulation of pituitary-adrenal responsiveness and individual differences in behavior of young domestic goats. *Physiology & Behavior*, *43*(4), 451–458.

McRae, K., Misra, S., Prasad, A. K., Pereira, S. C., & Gross, J. J. (2012). Bottom-up and top-down emotion generation: Implications for emotion regulation. *Social Cognitive and Affective Neuroscience*, 7(3), 253–262. https://doi.org/10.1093/scan/nsq103.

Meaney, M. J. (2010). Epigenetics and the biological definition of gene × environment interactions. *Child Development*, *81*(1), 41–79. https://doi.org/10.1111/j.1467-8624.2009.01381.x.

Miceli, S., Nadif Kasri, N., Joosten, J., Huang, C., Kepser, L., Proville, R., … Schubert, D. (2017). Reduced inhibition within layer IV of sert knockout rat barrel cortex is associated with faster sensory integration. *Cerebral Cortex*, *27*(2), 933–949. https://doi.org/10.1093/cercor/bhx016.

Munafò, M. R., Brown, S. M., & Hariri, A. R. (2008). Serotonin transporter (5-HTTLPR) genotype and amygdala activation: A meta-analysis. *Biological Psychiatry*, *63*, 852–857.

Nietzer, S. L., Bonn, M., Jansen, F., Heiming, R. S., Lewejohann, L., Sachser, N., … Schmitt, A. G. (2011). Serotonin transporter knockout and repeated social defeat stress: Impact on neuronal morphology and plasticity in limbic brain areas. *Behavioural Brain Research*, *220*(1), 42–54. https://doi.org/10.1016/j.bbr.2011.01.011.

Offenbaecher, M., Bondy, B., de Jonge, S., Glatzeder, K., Kruger, M., Schoeps, P., & Ackenheil, M. (1999). Possible association of fibromyalgia with a polymorphism in the serotonin transporter gene regulatory region. *Arthritis and Rheumatism*, *42*(11), 2482–2488. https://doi.org/10.1002/1529-0131(199911)42:11<2482::aid-anr27>3.0.co;2-b.

Pang, R. D., Wang, Z., Klosinski, L. P., Guo, Y., Herman, D. H., Celikel, T., … Holschneider, D. P. (2011). Mapping functional brain activation using [14C]-iodoantipyrine in male serotonin transporter knockout mice. *PLoS One*, *6*(8), e23869. https://doi.org/10.1371/journal.pone.0023869.

Pluess, M. (2015). Individual differences in environmental sensitivity. *Child Development Perspectives*, *9*(3), 138–143. https://doi.org/10.1111/cdep.12120.

Pluess, M., & Belsky, J. (2011). Prenatal programming of postnatal plasticity? *Development and Psychopathology*, *23*(1), 29–38. https://doi.org/10.1017/S0954579410000623.

Pluess, M., & Belsky, J. (2013). Vantage sensitivity: Individual differences in response to positive experiences. *Psychological Bulletin*, *139*(4), 901–916. https://doi.org/10.1037/a0030196.

Pluess, M., Belsky, J., Way, B. M., & Taylor, S. E. (2010). 5-HTTLPR moderates effects of current life events on neuroticism: Differential susceptibility to environmental influences. *Progress in Neuro-Psychopharmacology & Biological Psychiatry*, *34*(6), 1070–1074.

Pluess, M., Velders, F. P., Belsky, J., van IJzendoorn, M. H., Bakermans-Kranenburg, M. J., Jaddoe, V. W., & Tiemeier, H. (2011). Serotonin transporter polymorphism moderates effects of prenatal maternal anxiety on infant negative emotionality. *Biological Psychiatry*, *69*(6), 520–525. https://doi.org/10.1016/j.biopsych.2010.10.006.

Propper, C., Willoughby, M., Halpern, C. T., Carbone, M. A., & Cox, M. (2007). Parenting quality, DRD4, and the prediction of externalizing and internalizing behaviors in early childhood. *Developmental Psychobiology*, *49*(6), 619–632.

Risch, N., Herrell, R., Lehner, T., Liang, K.Y., Eaves, L., Hoh, J., Griem, A., Kovacs, M., Ott, J., & Merikangas, K. R. (2009). Interaction between the serotonin transporter gene (5-HTTLPR), stressful life events, and risk of depression: A meta-analysis. *JAMA*, *301*(23), 2462–2471.

Roiser, J. P., Rogers, R. D., Cook, L. J., & Sahakian, B. J. (2006). The effect of polymorphism at the serotonin transporter gene on decision-making, memory and executive function in ecstasy users and controls. *Psychopharmacology*, *188*(2), 213–227. https://doi.org/10.1007/s00213-006-0495-z.

Schipper, P., Hiemstra, M., Bosch, K., Nieuwenhuis, D., Adinolfi, A., Glotzbach, S., Borghans, B., Lopresto, D., Fernández, G., Klumpers, F., Hermans, E. J., Roelofs, K., Henckens, M.J.A.G., & Homberg, J. R. (2019). The association between serotonin transporter availability and the neural correlates of fear bradycardia. *Proceedings of the National Academy of Sciences of the United States of America*, *116*(51), 25941–25947.

Schjolden, J., & Winberg, S. (2007). Genetically determined variation in stress responsiveness in rainbow trout: Behavior and neurobiology. *Brain, Behavior and Evolution*, *70*(4), 227–238. https://doi.org/10.1159/000105486.

Schoebi, D., Way, B. M., Karney, B. R., & Bradbury, T. N. (2012). Genetic moderation of sensitivity to positive and negative affect in marriage. *Emotion*, *12*(2), 208–212. https://doi.org/10.1037/a0026067.

Sharpley, C. F., Christie, D. R. H., Bitsika, V., Andronicos, N. M., Agnew, L. L., & McMillan, M. E. (2017). Does psychological resilience buffer against the link between the 5-HTTLPR polymorphism and depression following stress. *Physiology & Behavior*, *180*, 53–59.

Sharpley, C. F., Palanisamy, S. K., Glyde, N., Dillingham, P.W., & Agnew, L. L. (2014). An update on the interaction between the serotonin transporter promoter variant (5-HTTLPR), stress and depression, plus an exploration of non-confirming findings. *Behavioural Brain Research*, *273*, 89–105. https://doi.org/10.1016/j.bbr.2014.07.030.

Sih, A., Bell, A., & Johnson, J. C. (2004). Behavioral syndromes: An ecological and evolutionary overview. *Trends in Ecology & Evolution*, *19*(7), 372–378. https://doi.org/10.1016/j.tree.2004.04.009.

Strathearn, L., Li, J., Fonagy, P., & Montague, P. R. (2008). What's in a smile? Maternal brain responses to infant facial cues. *Pediatrics*, *122*(1), 40–51. https://doi.org/10.1542/peds.2007-1566.

Suomi, S. J. (2006). Risk, resilience, and gene x environment interactions in rhesus monkeys. *Annals of the New York Academy of Sciences*, *1094*, 52–62.

Szyf, M. (2015). Epigenetics, a key for unlocking complex CNS disorders? Therapeutic implications. *European Neuropsychopharmacology*, *25*(5), 682–702. https://doi.org/10.1016/j.euroneuro.2014.01.009.

Todd, R. M., Ehlers, M. R., Müller, D. J., Robertson, A., Palombo, D. J., Freeman, N., Levine, B., & Anderson, A. K. (2015). Neurogenetic variations in norepinephrine availability enhance perceptual vividness. *The Journal of Neuroscience*, *35*, 6506–6516.

Troisi, A., Frazzetto, G., Carola, V., Di Lorenzo, G., Coviello, M., D'Amato, F. R., Moles, A., Siracusano, A., & Gross, C. (2011). Social hedonic capacity is associated with the A118G polymorphism of the mu-opioid receptor gene (OPRM1) in adult healthy volunteers and psychiatric patients. *Society for Neuroscience*, *6*(1), 88–97.

van der Doelen, R. H., Deschamps, W., D'Annibale, C., Peeters, D., Wevers, R. A., Zelena, D., … Kozicz, T. (2014). Early life adversity and serotonin transporter gene variation interact at the level of the adrenal gland to affect the adult hypothalamo-pituitary-adrenal axis. *Translational Psychiatry*, *4*, e409. https://doi.org/10.1038/tp.2014.57.

van Ijzendoorn, M. H., & Bakermans-Kranenburg, M. J. (2015). Genetic differential susceptibility on trial: Meta-analytic support from randomized controlled experiments. *Development and Psychopathology*, *27*(1), 151–162.

van Ijzendoorn, M. H., Belsky, J., & Bakermans-Kranenburg, M. J. (2012). Serotonin transporter genotype 5HTTLPR as a marker of differential susceptibility? A meta-analysis of child and adolescent gene-by-environment studies. *Translational Psychiatry, 2*, e147. https://doi.org/10.1038/tp.2012.73.

Volf, N.V., Kulikov, A.V., Bortsov, C. U., & Popova, N. K. (2009). Association of verbal and figural creative achievement with polymorphism in the human serotonin transporter gene. *Neuroscience Letters, 463*(2), 154–157. https://doi.org/10.1016/j.neulet.2009.07.070.

Wills, T. A., Vaccaro, D., & McNamara, G. (1992). The role of life events, family support, and competence in adolescent substance use—A test of vulnerability and protective factors. *American Journal of Community Psychology, 20*(3), 349–374. https://doi.org/10.1007/bf00937914.

Wolf, M., van Doorn, G. S., & Weissing, F. J. (2008). Evolutionary emergence of responsive and unresponsive personalities. *Proceedings of the National Academy of Sciences of the United States of America, 105*(41), 15825–15830. https://doi.org/10.1073/pnas.0805473105.

CHAPTER 6

Clinical assessment of sensory processing sensitivity

Elaine N. Aron
Department of Psychology, Stony Brook University, Stony Brook, NY, United States

Contents

1 Assessment of highly sensitive persons 137
  1.1 A blend of approaches to assessment 137
  1.2 When other seemingly contradictory traits are present 139
  1.3 Taking into account differential susceptibility 140
2 Working with HSPs 142
  2.1 When the therapist is not highly sensitive 143
  2.2 When both therapist and client are highly sensitive 145
  2.3 Five goals for these clients 146
  2.4 Gender, culture, ethnicity 147
3 Typical issues working HSP issues in a clinical context 148
  3.1 Overstimulation 148
  3.2 Emotional regulation 152
4 Personal relationships, career, and the workplace 156
  4.1 Research on relationships 157
  4.2 Adjusting to those who are not as sensitive 158
  4.3 Getting off to the right start 160
5 Conclusion 162
References 163

While the *The Highly Sensitive Brain* is not written specifically for clinicians, the clinical context provides another view of sensory processing sensitivity (SPS). Both the value and the drawback of such a chapter are the somewhat subjective nature of what is here, either impressions gained from more than 27 years interviewing, teaching, or working in therapy with the highly sensitive, or attempts to generalize from the broader research on SPS. Regarding some of the clinical research, specifically, this was discussed at length in previous chapters. In sum, some studies have pointed to a general vulnerability to anxiety, depression, and burnout in association with SPS (Greven et al., 2019). However, the conclusions are somewhat questionable without data on the quality of childhood experience and the external stressors that highly sensitive individuals may be facing. This is due to the

The Highly Sensitive Brain
https://doi.org/10.1016/B978-0-12-818251-2.00006-0

strong evidence backing the theories of differential susceptibility (Belsky & Pluess, 2009) and vantage sensitivity (Pluess & Belsky, 2013). It appears that some individuals, even if highly sensitive, may be well adjusted, and they may not necessarily suffer from depression, anxiety, or burn out if they have had high-quality childhoods and continue to be in nurturing and positive environments. On the other hand, when exposed to harsh, negative, and unsupportive environments, especially in childhood, the highly sensitive tend to express many of the negative outcomes that were discussed in previous chapters, including stress, depression, and anxiety; and lower satisfaction with life, work, and personal relationships. Hence, generalizing these results to all highly sensitive persons (HSPs) is very questionable. One also has to wonder how results might have differed, especially in studies of burnout, if those in the sample had known about their trait and adjusted their lives accordingly.

For those readers who are clinicians, whom I will often address as "you," this chapter will have obvious benefits. Being knowledgeable about SPS in the clinical context offers a significant, special opportunity to be helpful to clients, sometimes extraordinarily so. HSPs who do not know about their sensitivity often remark to me, "This has changed my life." Some have claimed it has ended years of depression and even ended suicidal thinking. These are the kinds of rewarding changes we would all like to see, and often it is accomplished in one session. It is sad for both therapists and clients that there are so few therapists with training or knowledge even of innate temperament in general, much less SPS, when it is found in 20%–30% of the population (Aron & Aron, 1997; Pluess et al., 2018), and undoubtedly there is a higher percentage in therapy. As such, it is important to educate clinicians (and individuals) about SPS so that they may facilitate the thriving of highly sensitive individuals.

This chapter covers four topics: assessment of one individual for clinical purposes (as opposed to in research settings, see Chapters 2 and 4); specific considerations when working with HSPs; helping HSPs with their three most common problems (being easily overstimulated, their strong emotional responses, and their low self-esteem); and finally, the issues arising out in the world, in relationships and work.

The chapter does not discuss specific treatments, such as cognitive behavioral or depth psychotherapy (even though both, and others, are evidence-based). I assume most of you already prefer certain methods and are proficient in them. I also do not explore other modalities that may be quite helpful for highly sensitive people (e.g., yoga, meditation, certain types of exercise) because they are noted in previous chapters of this volume and the list would be incomplete, thus implying that those on the list are the best, or good for all sensitive clients, when individual factors may be decisive. Also, when it comes to any wellness or self-care routine, it is our opinion that adherence and personal preference is key, so that if a client likes something, they will do it again; while even if we tell them something is "good" for them, but they do not enjoy it, the probability seems low that they will stay with it. You may well want to recommend such activities and programs, however, and then see how they work out for individual clients (your own research).

1 Assessment of highly sensitive persons

Assessment can be approached in various ways, but one must be attuned to how the information will be assimilated by the client. Readers of *The Highly Sensitive Person* have often commented that they cried much of the time while reading it, and either devoured it in one sitting or had to space the reading over months. Although therapists will hopefully become habituated to the term and its implications, clients new to it may feel as if they are "coming out." They have been keeping secret all these separate embarrassing problems without ever knowing the scope or the value of what they were hiding or that there were others like them. Having the complete picture can be quite profound.

Particularly men, women from minority groups (raised always to be "strong"), and those in certain professions (e.g., the armed forces) may find the idea that they might be highly sensitive very disturbing. In that case, you, as the therapist, can keep the idea to yourself as a hypothesis and gather information, focusing on what might be the person's enhanced abilities resulting from their trait. For example, a policewoman struggled with the idea of being highly sensitive until she thought about her unusual sense on the street of "what's coming down" that other officers often lacked. Similarly, a soldier on active duty recalled his suppression of his sometimes overpowering feelings by maintaining an almost obsessive and successful focus on the subtle strategies and details (the tell-tale sign) that would keep his unit successful, but also himself and his unit safe.

1.1 A blend of approaches to assessment

Psychologists generally assume that personality, being that it is a blend of history and temperament, is difficult to assess separately from "a trait such as" SPS. In fact, it is not very difficult. The solution comes from a blend of approaches. One is client spontaneous self-report, requiring the therapist only to be familiar with the trait and mentally record reports of sensitivity indicators such as those found on the HSP Scale (Aron & Aron, 1997). For example, one may observe "high sensitivity" when a client is suffering acutely from the overstimulation of having a job and a young child, and comparing herself to the "super moms" around her.

A highly sensitive (HS) man complains of the aggressively competitive environment in his office or his witnessing unethical behaviors, which HSPs generally find disturbing. A HS father is overwhelmed by witnessing the birth of his child or having to hear his infant's crying. On their own, these self-reported problems do not signal SPS, but as they add up in a client, one becomes more certain.

The HSP Scale is also helpful. You may ask about or watch for items on the scale, since it is research-based, and while not designed for individual clinical assessment, it is reasonably accurate and widely used by researchers investigating SPS. For example, you might hear or ask about an item such as the next to the last, about being made nervous while competing or being observed. Perhaps your client is discussing having been very nervous or failing when being observed at work by a manager or coworkers, taking a timed test, competing at the next level in a sport, or performing before a large audience, even though they are perfectly competent when doing these things when not under these conditions.

When it seems appropriate, perhaps after broaching the subject in a small way, you might give the scale or causally mention that they might be interested in taking the self-test at HSperson.com or download a smartphone-based application (the LoveSmart app) that allows them to assess their sensitivity (using the HSP Scale) and provides feedback on their romantic relationship.

You may also observe or ask questions while keeping the four aspects of SPS (which are assessed in research), considering whether you have noticed (1) depth of processing in the questions they ask, for example, their expressing a need for meaningful work; (2) being easily overstimulated compared to others when, for example, at a mall or in large crowds; (3) having strong emotional responses in that they cry or blush easily or show unusual empathy for those treated unfairly or even concern about your well-being; and (4) noticing subtleties, maybe even while with you, such as a change in the arrangement of the furniture or your being a bit hoarse today.

Clues are also found in history, for example, the client simply saying something similar to the last item of the self-test, "Parents or teachers seemed to see me as sensitive or shy." Notice "shy" is used in this item because many sensitive children are seen as shy when they are in fact only pausing to observe before acting, or have become shy (technically afraid of social judgment) because of the trait underlying their shyness. They have felt different or perhaps expressed themselves poorly in social situations because of being overstimulated, and when this repeats, they fear social judgment for it even before it has happens—a common feedback loop. They may remember crying easily or being more vulnerable to pain, and boys especially may relate how they were especially teased for it.

Overall, they may seem to have been more troubled by what seem to be typical life events than would be more "resilient" children. They may describe a seemingly small experience, such as being lost for a few minutes in a store, in such a way that you and they feel it was traumatic and still affects them. This is due to their differential susceptibility and its underlying cause, the strength of their emotional responses, both positive and negative. Their integrative processing of the details of a stimuli or situation, as shown by brain research (Acevedo, Aron, Fisher, & Brown, 2012; Acevedo, Aron, Pospos, & Jessens, 2017), is associated with memories, both positive and negative, that guide behavior when facing similar subsequent stimuli. However, for the highly sensitive, the impressions left in the brain's architecture can be especially strong due to the processes described in Chapter 5 on the neurobiology of SPS. While these memories may have led to more life difficulties, their memories also may make the sources of their difficulties more accessible. As such, the highly sensitive may be better equipped to "learn from mistakes," exercise caution in the face of a "known" threat, and engage in creative strategies for staying safe from harm and avoiding potential stressors.

1.2 When other seemingly contradictory traits are present

Assessment is probably most complicated by the mixture of other, seemingly opposite traits, specifically extraversion and high sensation-seeking. Research (e.g., Aron & Aron, 1997) has found only a moderate correlation of SPS with introversion (about 30% appear to be extraverts). These extraverted sensitive clients may have many friends, enjoy meeting new people, and speak up in social situations. They differ from other extraverts in that they need "downtime" just as much as an introverted sensitive person. That is, they need equal time to process their experiences and recover from the stimulation of the day. But more than introverted sensitive clients, they may express their thoughts in social situations, which may often be unusually creative or deep, and then find themselves having to defend their ideas in the face of doubting or hostile skeptics, which they then find distressing.

Similarly, high sensation-seeking is a trait orthogonal to SPS, according to our unpublished research. While it may seem it would be the opposite, the opposite of SPS is actually impulsivity. HSPs with this trait of high sensation-seeking speak of "one foot on the gas, one foot on the brake." They might sound self-defeating in their engaging in situations that turn out to be "too much." The cause in these cases, however, is actually simply the combination of the two traits. You may hear of relationships or jobs in which the client found the needs of one of the two traits well satisfied and the other not. (An example of a good career for them is free-lance journalism; ideally a relationship would be with someone else having both traits.) High sensation-seeking, highly sensitive types will be experts on safety if they have chosen to engage in something that seems risky, and are equally likely to be introverts, who for example rock climb alone or love foreign travel, to the most foreign place possible, but declare they would never travel in a group tour or on a cruise ship.

1.3 Taking into account differential susceptibility

Highly sensitive clients will vary in the intensity of their presenting symptoms, and also how their trait is expressed and visible through their immediate presentation. Some may be very high functioning individuals who simply know nothing about their trait and are troubled because of their feeling more highly stressed than others in the same situation, or finally wondering about any psychological contribution to their many somatic problems (e.g., Irritable Bowel Syndrome, chronic muscle tension). They may want to talk about their private "over reaction" to criticism at work, or what they see as too much difficulty with a major decision.

A client may also have a general, underlying sense that there is something wrong with them, often beginning in childhood. Sometimes you suspect that they are overcompensating for this sense of an innate flaw through exceptional achievements in their career, academic work, or physical prowess. But the history of these high-functioning sensitive clients will sound fairly positive, especially if their parents understood, accepted, or even liked and encouraged their child's manifestations of high sensitivity.

Others, at the low end of emotional or social functioning due to differential susceptibility, may be so affected by a negative childhood and specific traumas that their sensitivity is uncovered only by looking between the lines, that is looking beyond the overt symptoms. Sometimes you hear about how they were seen as children quite positively by more impartial adults or friends and away from trauma, as if you and they are bracketing the history to find the rest of the story. If you are not certain, stay with the rest of the assessment and your treatment plan. It will not matter if your discussion of their trait is delayed until you both can see it. At that later time, it may be a greater relief to the client to understand this piece of their story than if you introduce it early, as if explaining away their pain from childhood.

It helps to gain a sense of the "pure" trait without so many symptoms by thinking about people you know who seem to be both highly sensitive and functioning well, even exceptionally well. The only sign of their sensitivity may be their taking more time for "solitude" than others. Otherwise one may simply know them as highly creative, especially empathic, ethical, spiritual, good at business, insightful about world events, exceptionally conscientious, etc., without associating these individual qualities with high sensitivity. Persons in history or in the news who function well but may be sensitive are also helpful for developing a sense of the trait uncomplicated by immediate, obvious symptoms. Although I try to avoid declaring someone as highly sensitive without their permission, I have ventured to guess the U.S. President Abraham Lincoln might have been. With this perspective on the trait, aspects of it will emerge in the more troubled client too.

Another person I cannot resist thinking of as highly sensitive is Etty Hillesum, based on her diaries from 1941 to 1943 (Hillesum, 1983). She was a Jewish woman who grew up in Amsterdam in the years preceding the Second World War. She described her parents and her childhood in glowing terms, suggesting she was on the high end of functioning with her sensitivity. She was reaching adulthood when the Nazis came to power, witnessed the Holocaust as it unfolded in Holland, and finally died in Auschwitz. Her diaries testify to how a HSP with the right background might manage exceptionally well, both the overwhelming levels of stimulation (e.g., constant danger, crowded living conditions, near starvation) and a situation evoking the most intense emotions. Her sensitivity is clearer from the entire body of her work, but on November 10, 1941, for example, she wrote: "Mortal fear in every fiber. Complete collapse. Lack of self-confidence. Aversion Panic" (p. 58). She used all the usual methods of affect regulation, from distraction, seeking comfort from others, and withdrawal for time alone, to gratitude for one's blessings, helping others, and seeking a positive meaning in negative events (Larsen & Prizmic, 2004). Hence, she also could write on July 3, 1942: "Very well then, this new certainty, that what they are after is our total destruction, I accept it… I shall not burden the others with my fears…

I shall not be bitter if others fail to grasp what is happening... I work and continue to live with the same conviction and to find life meaningful—yes, meaningful" (p. 160). In the end, she helped others with her selfless kindness, partly inspired it seems by her "big picture" of history. (In as yet unpublished data, highly sensitive subjects agree significantly more often with the item, "Liking to imagine things in the past," as well as "being willing to sit at the bedside of a dying person and comfort them").

As for your final decision on a client's level of sensitivity, realize that there will inevitably be some false positives. That is, clients that you will think are highly sensitive, but who are actually not. Often, they persuade you of it or bring in a score to prove it, in the end hoping the two of you can establish that their distress or impairments are not indicative of any problem but entirely due to their sensitivity. Slowly you set them right.

There will also be false negatives—persons you thought were *not* highly sensitive but actually were. Hopefully you will make fewer of these errors than those therapists who see highly sensitive clients as impaired due, for example, to being easily overstimulated by noisy restaurants, an odor where they work that seems toxic only to them, difficulty making decisions, needing time away from others, and so forth. Without knowing that there are in fact a significant number of people with the trait, even if they are a minority, and not understanding its evolutionary origins and its advantages, these complaints could indeed seem to be due to an anxiety or personality disorder. They would seem to be "associated with present distress...or disability... or an important loss of freedom."

Finally, perhaps the most important category will be all those who are both highly sensitive and also have a true, diagnosable disorder. The art of distinguishing the trait from other syndromes is gradually developed through experience, but a summary of my own experience is found in Appendix B of *Psychotherapy and the Highly Sensitive* Person (Aron et al., 2010), "Distinguishing Sensitivity from DSM Disorders," (based on the previous DSM but still helpful) which includes how co-occurrences may differ in their presentation.

2 Working with HSPs

The most important quality and attitude you must bring to work with the highly sensitive, in my experience, is kindness. Kindness underlies the other forthcoming points. For example, it is the basis for helping the client accept the trait as a fundamentally positive aspect of their nature, or maintaining a comfortable level of arousal for them during sessions. Remember that these clients process deeply whatever you say, and with their typical low self-esteem will process it more negatively. Due to differential susceptibility, which, again, involves a high attunement to both positive and negative interactions, what they can hear as genuinely positive will have particularly profound effects. Indeed, the success of your work with a client may hinge on this. Although we do not have data on individual therapy, we do know that those scoring in the highest 30% on the HSP Scale gain more than low scorers (who essentially gain almost nothing) from mental health interventions to reduce depression (Pluess & Boniwell, 2015) and bullying (Nocentini, Menesini, & Pluess, 2018). That is, they are also differentially susceptible to benefiting from help and supportive environments, as discussed in previous chapters.

2.1 When the therapist is not highly sensitive

You may wonder what the effect is of not being highly sensitive yourself. There is no research on this as yet, but let's consider the advantage of not having the trait. Your respect and admiration is even more meaningful. You will easily role model how to turn up the "volume," the intensity of speech in its loudness, brusqueness, and bluntness. Sensitive clients need to raise their volume in order to be heard by others like you. Also, when you are inevitably too blunt sometimes or ask questions in a way that sensitive clients find threatening, you are providing the chance to learn how to handle such communications and find out what was the actual intention behind the volume.

Also, by being moderate or low in the sensitivity trait, you provide a close-up experience of how the majority think, experience them, and in general what can and cannot be expected from those without the trait. Since you may find their reluctance to try something new a bit vexing, you might not always be inclined to take "no" for an answer when you think a client can do something and will be better for it. Often these clients have experienced people giving up on them too soon, out of exasperation or overprotectiveness. Since the client wants to succeed, this respectful confidence, added to sympathetic support, can be extremely hopeful.

When sensitive clients are overwhelmed, you might be calmer yourself and thus more soothing than a highly sensitive therapist would be, and your clients will internalize this calmer response. Finally, since you probably tend to speak and act more freely, you provide a role model for being more spontaneous.

The main danger of not possessing the trait is a huge one: You may unconsciously persist in the common human assumption that others are like you, and that all of a client's difficulties or differences are due to a trauma that has to be uncovered and healed, or to a thought process that has to be corrected. Even if you do engage without such assumptions, the slowness of the entire process can be frustrating. You may not like having to do everything so cautiously. You will need patience, and your consolation is that "patience" is a good virtue to develop for working with everyone, but especially with the highly sensitive.

You may also experience boredom to a greater extent in sessions with HSPs than would a therapist with this trait. Sensitive clients may have deep insights, but these can surface slowly. When they are silent and you know they are thinking, it is tempting to probe for their thoughts or offer them multiple choices. "I notice you seem to be having trouble putting your thoughts into words—is this true? Is this more so than usual? Or might there be something especially difficult to say? Perhaps about our relationship, such as my upcoming vacation? Or am I interrupting some reflection?"

To stay attuned, you must try to turn down your thought process and "volume"—the forcefulness of your words. Everyone tends to put their thoughts into words with the same intensity as is required for others to get through to them. The highly sensitive, however, only need hints, gestures, glances, nuances, and tones of voice, and that is what they usually give out. The nonsensitive therapist's suggestions may seem like strong orders to be obeyed invariably ("Don't leave a message after nine at night" may only mean that you often do not check for messages after nine, but can be heard as don't be a nuisance and bother me after nine.). Questions can sound like attacks ("Why did you do that?" "What were you thinking when you did it?"). Interpretations can come down as judgments about the integrity of their intentions ("Maybe the real reason you bring me these apples from your tree and then the muffin you baked is that you're afraid I will reject you if you offer me more than just yourself.").

Especially when the highly sensitive are cautiously uncovering their subtle feelings, you must tread with great care. These feelings are like sea creatures brought up from great depths and barely able to survive in full sunlight. Your smile or digression may simply seem too irreverent, so that they will not share that type of material again, even though it is central to their lives. As one client put it, "It's as though I was in a dark, candlelit room praying, and then he came in, turned on the overhead light, and asked me what I was doing."

Of course, these clients can learn to adapt to a nonsensitive volume—many have already—and may gain so much from that, or from you in other ways, that it ceases to trouble them. But one danger with high volumes is that you will be having too much influence even though you intend to develop the client's independence. The client may say nothing about your influence (and of course you want to have some), and possibly not even know about it, being so used to being overruled.

All in all, you may find you have unintentionally hurt a sensitive client's feelings or disappointed them through your insensitivity, possibly finding out only when the client leaves treatment. It is helpful to check in gently but often to see how the work and the relationship are progressing in their eyes, even asking specifically a few times about your own lack of the trait and mutually exploring the advantages as well as the disadvantages of that.

2.2 When both therapist and client are highly sensitive

The advantages are obvious. The two understand each other immediately. The client feels comfortable, well heard, and has the role model of another sensitive person who is coping relatively well in life. The therapist has real experience with over arousal, intense emotions, and setting boundaries. The therapist is not thrown off by or made envious of the depth and wide-ranging intuition of the client. Above all, just by being themselves, sensitive therapists improve the self-esteem of their sensitive clients.

However, even when sensitive therapists are at home with their trait, there are a few pitfalls to avoid. One is assuming the client is similar in other ways as well. For example, I tend to expect highly sensitive clients to be introverts like me, but some are not. I may ask them to remind me when I am assuming it, and I have to adapt myself to sensitive extraverts' ways of thinking. I may also assume a sensitive client has the same limitations and preferences that I have. I am careful to hide my dismay if a sensitive client takes up hang-gliding or aims to run for public office, goals I would neither have nor feel I could achieve.

In addition, when discussing the client's other relationships, especially marriage to a person without the trait, you and the client can inadvertently forget that person's perspective. It helps if highly sensitive therapists also have significant others in their lives who are not highly sensitive.

Sensitive therapists may also have issues that would impact all clients, but perhaps sensitive ones more. For example, having been the target of negative stereotyping yourself, you may have come to feel flawed and tend to submit in a conflict. Thus, you may have difficulty being assertive about your needs, such as proper compensation or undisturbed down time. You need to call on your conscientiousness and remember that these behaviors are role-modeling poor boundaries, the last thing sensitive clients need. Watch whether you postpone a difficult confrontation, take the blame too quickly in an interaction, or even feel deeply guilty or ashamed at times, all without considering whether these feelings might be due to unconscious pressure from the client that you submit to their will rather than simply your empathy or kindness. As a HSP yourself, you can understand their needs better, but that does not mean you have to fulfill them.

Adapting to the needs of the sensitive client can lead to a satisfying professional success, but if it requires that you first work on your own issues about your sensitivity, you have the opportunity for a satisfying personal success as well.

2.3 Five goals for these clients

One way to conceptualize what your highly sensitive clients need is to focus on these five goals. First, they need to feel certain that this trait is real. Many will doubt it, at least partially. Or they will see it only as a disadvantage, whereas a full understanding, through you, reveals its obvious advantages. For some it will be important that you refer to the research on brain activation differences, as found in other chapters—something concrete and relatively objective.

Second, your clients need to reframe their life in the light of their high sensitivity. Many events that left them feeling inferior, such as perceived failures in overstimulating social situations or during performances that were important to them, were due to their sensitivity. Or they may need to revisit decisions they later regretted, for example, declining a job that involved frequent travel or supervising a large number of people. This reframing can be done as the memories come up, but you do not want to forget to look at every low-self-esteem-inducing situation being discussed as possibly influenced, sometimes decisively, by the client's trait.

Third, they need to heal from past traumas, relationship or work failures, and whatever else has brought them to you. That is your shared primary goal. Since this work often involves depression or anxiety, medication may become an issue. If it seems that your sensitive client needs medication, my experience and those of psychiatrists familiar with the trait amply demonstrates that they should begin on a very low dose, as they are very susceptible to side effects, which may cause them to want to stop. Also, a low dose is often all that is needed to see improvements. If they are reluctant, but live with people they care deeply about, it can help to remind them how affected these others are by their depression, anxiety, etc.

Fourth, they need to develop a lifestyle that suits a HSP. Usually they have copied the habits of the majority, who do not have this trait, although this lessens with age as even those not knowing about their sensitivity may begin to adopt a more suitable lifestyle. Lifestyle is a particular problem for those under 30, who are still trying to fit in with their generation as well as still having the stamina for handling high levels of stress from multitasking, lack of sleep, trying to impress teachers or employers, and, all at the same time striving to create meaningful personal relationships and possibly begin a family. Given all of this, frequently those around 30 come to therapy specifically due to the results of their lifestyle not working for them. A more appropriate lifestyle will be discussed here.

Finally, fifth, it helps highly sensitive individuals to know others with their trait, and to be in a group of entirely highly sensitive people at least once, as it is a unique experience. There are more and more opportunities to do this. However, if there is a therapist involved, perhaps facilitating a support group, it's important for the therapist leading the group to have experience working with the highly sensitive, diagnosing comorbid issues, and being able to speak about helpful, adaptive versus maladaptive emotional management strategies.

2.4 Gender, culture, ethnicity

A therapist's gender and cultural sensitivity is as important with these clients as with those who are not so sensitive, but the issues may differ, especially with sensitive men and minorities.

It is crucial to appreciate the difficulties faced by highly sensitive men, depending on their culture. Cultures established by relatively recent immigrants, North and South America plus Australia and New Zealand, tend to see sensitivity in a man as a problem, a sign of weakness, or associate it with homosexuality. Using interviews, Zeff (2010) compared five cultures: Canada, U.S., Denmark, Thailand, and India, and he found large differences in how sensitive men were viewed. For example, in comparison to the first two countries, in the last three countries, the men reported that as boys they were rarely, if ever, teased for their sensitivity or lacking in friends; although almost all, whatever the country, felt there was something wrong with them. Parents, especially fathers, try to reduce its possible, but not inevitable, manifestations (e.g., crying easily if this has not been suppressed, pain sensitivity, hesitation to enter new situations without the opportunity to observe, lack of enthusiasm about competition in comparison to strong artistic, academic, or social justice interests). Highly sensitive men's reaction to realizing that they are HSPs is often highly variable—which may result in either intense relief and gratitude, or distress and denial. Helping them take pride in their trait and trust it is an important part of therapy.

Minorities (e.g., African-American, Asian, Muslim, Latino, gay, lesbian) have a double issue, in that they have suffered stereotyping and prejudice already, usually wounding them deeply, and their high sensitivity narrows their self-identity even more if they allow it. African-Americans and recent immigrants in particular have been convinced they must be strong. Anything less is unacceptable. Nonetheless, I have observed that highly sensitive African-American men in particular have the best chance of leaving behind dangerous environments they may have grown up in. There are many reasons why this might be the case. With all minorities, it is probably wiser to develop an authentic understanding of their original painful experience as a member of a minority before introducing this new element of their sensitivity, as the client may perceive that to be your own agenda.

3 Typical issues working HSP issues in a clinical context

The only inherent problem with being highly sensitive is being easily overstimulated. Emotional regulation, the second issue to be discussed, is a result of high emotional responsiveness, which is often an advantage as well, and probably evolved as an aid to depth of processing: We learn better when motivated more strongly (Baumeister, Vohs, DeWall, & Zhang, 2007).

Low self-esteem is also a problem, not inherent to SPS, but arises from personal history and culture. In my unpublished data, the highly sensitive were significantly less likely to answer "yes" to the single question, "Do you feel good about yourself?" This was true even when measures of negative affect and childhood were partialled out. There are data to suggest that a large contribution to this not feeling good about one's self is cultural. In China, 20 years ago at least, sensitive, quiet children were among the most popular, and in Canada among the least popular (Chen, Rubin, & Sun, 1992; although this has changed in China; Chen, Cen, Li, & He, 2005).

3.1 Overstimulation

Chronic overarousal due to overstimulation is often the presenting problem—"feeling stressed out," "suffering from burn out," feeling "overwhelmed," or a sense that "I can't handle things anymore." This will affect restfulness and health; and with time, performance on the job or with close others such as family. Usually the highly sensitive allow their own duties to be affected last—part of the problem. Hence, they are in therapy because they see no options and think there is something wrong with them that they cannot handle as much as others.

Overarousal often occurs for the highly sensitive even more than others during important life transitions, as these usually require rapid shifts in stimuli, foci of cognition, and behaviors, while they, by nature, would prefer to go slowly and reflect on what is happening. Even pleasant changes such as leaving on a vacation, moving to a new home, marrying, getting a promotion, parenthood, or retirement will have unpleasant side effects, such as feeling unsettled, losing sleep, and feeling flooded by their emotional reactions. Often the problem can be observed in your office, at the start and end of sessions, or the start and ending of a longer separation from the therapist. (This would be over and above separation distress due to personal history.)

When clients complain of panic attacks, they can be the result of an experience of intense overstimulation, leading to a terrifying sense of not being able to cope and fear of such panic happening again. These differ from more typical panic attacks in that they clear up quickly when their cause is explained and a few remedial measures suggested.

Another behavioral effect of overarousal is that it is simply avoided, whether knowingly or not. For many, the unpleasant feeling of being overaroused is associated with feelings of failure and humiliation. It is also very difficult to distinguish overarousal from the physiological experience of fear. Many presenting problems can be related to avoiding over stimulation and over arousal. For example, trouble making new friends may be due to never going to places where there may be strangers.

Maintaining optimal level of arousal in the office is a good starting place and that begins with the space itself. You may choose to make it an uncluttered environment with restful décor, for example. I try to dress in a way that does not attract attention and I prefer no music in the waiting room, but you may choose to reduce different sources of stimulation.

A central reason to avoid unnecessary overstimulation in the therapeutic context is that suggestions or interpretations made when a client is overaroused are often forgotten, and these sessions will usually be unpleasant for the client and perhaps even retraumatizing. However, maintaining an optimal level also does not mean that overarousal should never occur. Overarousal and overwhelming emotions are almost inevitable in effective psychotherapy, especially when trauma or defensiveness are involved, and it can be invaluable for seeing where the work should be focused. It also provides the opportunity to see how readily and in what situations the client is overwhelmed and to help the client regulate these overwhelming affects. But, again, in general the client will gain more from a session if you can prevent overarousal.

Some highly sensitive clients will learn to say when they want the therapist to slow down. Others will learn, consciously or unconsciously, to signal this, even if they cannot say it. I have had clients develop dizziness, a headache, nausea, or even blurry vision. If they come late or skip sessions or keep to small talk, you might suspect overarousal in other sessions as the cause.

However, when overarousal occurs, there are many ways to calm a client. You can drift off into something a bit mundane or connected to the outer world. Sensitive clients may well know what you are doing, but you can discuss later what you did and why. You might ask them to notice what is going on in their bodies. This may intensify the problem, but if it works, will not require leaving the topic altogether. The same is true of talking about the overarousal itself and why it may have come up, using the past tense to suggest it is going or gone.

When arousal is back in the optimum range, in that session or later, it could help to acknowledge what caused the sense of being overwhelmed and compare this to prior experiences. One can also encourage speaking up in the session when it occurs, or just taking a deep breath. One goal is to add to the client's repertoire of affect and arousal regulation.

One can think of the problem of being easily overstimulated outside of the office in two ways: short term, that is coping with over arousal before, during, and after it occurs; and long term, having the appropriate lifestyle. Often the second permits the first, however. For example, in the short run, good self-care such as adequate sleep, healthy diet, exercise, and "downtime" such as meditation or time in nature all help cope with stimulation in the short run, but one must have a lifestyle that allows time for these.

There is a great deal written for the public in *The Highly Sensitive Person*, the hsperson.com blog, and other places about the short-term problem of daily avoidance of overstimulation through adequate self-care, such as the measures suggested here. Besides concrete suggestions, you want to watch for a client being overall "too in" as well as "too out there." Too much self-care can arise from a fear of overstimulation so intense that it must be avoided at all costs.

You may find that how clients were taken care of in childhood greatly influences how they take care of themselves. Those who were neglected will often neglect themselves. Those who were praised for being hardy or for not making trouble may suppress their experience of overstimulation. Those whose parents had no idea about sensitivity maybe be unaware of the trait, or attribute their temperament and response to things as being high strung, high maintenance, or lacking resilience. On the other hand, those who obsess over their own sensitivity were usually overprotected or fussed over as children. Somatic complaints are common in those who were abused, afraid to speak up about their misery from overstimulation (at home or at school), or who learned to be quiet about physical symptoms unless they were truly sick. Whatever the nuances of their original care, it will probably be reflected in current distortions and understanding these together can help to optimize self-care.

Long-term issues can generally be conceptualized as problems with boundaries. Can they ask or tell the other person in a close relationship that they need to stay home from something they know is too much for them and need some downtime or solitude? At work, can they refuse working long hours when they can prove it will affect not only their well-being but also productivity? Therapists may see great gains simply by teaching clients all the nuances in ways to say "no" and otherwise maintain the boundaries necessary for their well-being. For example, it may help to suggest to the client that when saying "no" there is no need to explain. "Traveling cross-country next Monday simply isn't going to work for me." Explanations often leave openings for the other person to try to solve the issue when the issue is really simply "I can't." Unless of course an explanation can open a fruitful negotiation: "It's just not going to work for me, although I am willing, if I can take a day off when I return."

Seeing the history of clients' boundary problems can help to resolve them. Generally, you and they find they were not allowed boundaries in childhood (e.g., being given choices; their opinions solicited on matters affecting them; certain places and belongings respected as belonging to them). Usually because of this, or general low self-esteem, they do not assert themselves as adults. After all, part of the privilege of higher rank and influence is deciding the boundaries (a private office, a longer vacation, what activities a pair will engage in, or the right just to say "I'm so busy right now—don't call me, I'll call you.")

Some believe they can only earn the right to boundaries through high achievement, burning themselves out as they overcompensate for what they see as a shameful weakness to be hidden. In these cases, they may be unable to set boundaries inside themselves, to self-regulate this driven part of themselves. The same is true of high sensation-seekers. In these cases, it helps to think of the highly active and the highly sensitive parts as needing to learn to negotiate for the well-being of the whole.

Finally, be alert to when problems of overstimulation cannot be avoided, due, for example, to taking care of small children or elderly parents, having no choice but to work in a highly stimulating environment, or not being able to afford a safe or quiet place to live. When poverty is the root cause of overstimulation, sometimes you may know of overlooked community or family resources. Sometimes, too, the creativity of the highly sensitive can be released and innovative solutions found when the therapist acknowledges the constraints and helps them to think in a supportive, realistic way.

3.2 Emotional regulation

Given that those high in SPS have stronger emotional reactions and greater empathy (another reason for poor boundaries), they must develop robust emotional regulation skills—having the right emotion, at the right time, with the right intensity. Sometimes HSPs are "emotional leaders," signaling to others through their own reactions the appropriate response (e.g., by being the first to cry, to express anger, to act compassionately). Of course, some, especially men, may have to learn to have more emotion, more often, more intensely, having suppressed their awareness of their feelings. But generally, clients would like greater control over emotional responses, where possible.

Research (Brindle, Moulding, Bakker, & Nedeljkovic, 2015) has found that the highly sensitive are more aware of and have more negative emotions—depression, anxiety, feeling very stressed—than other people. (However, positive emotions were not measured and in other research, as described in previous chapters, their scores on measures of positive affect have exceeded those without the trait.) Using a measure of emotional regulation and another of distress tolerance, researchers (Bakker & Moulding, 2012) found that those high in SPS have special difficulty with the following:

1. Acceptance of feelings.
2. Shame about negative feelings.
3. Belief they can cope as well as others do.
4. Trust that negative feelings will not last long.
5. Assuming they can do something about their negative feelings eventually.

These results are not surprising and are tightly linked. Given that those high in SPS tend to have stronger negative emotions, these indeed can be difficult to accept, both in themselves and in comparison to others. The latter could lead to shame about these feelings and doubt that they can cope as well as others. They may indeed find that their negative feelings last longer and, given their prevalence and duration, they may begin to feel hopeless about doing anything about them. In another article (Bakker & Moulding, 2012), measures of dispositional mindfulness and acceptance of feelings moderated the relationship between SPS and anxiety in particular. That depression was not reduced may be due the origin of depressive feelings in the client's history, including early insecure attachment. Still, all of these points leave openings for therapists to help clients tolerate and accept negative states: Teaching emotional regulation skills, such as seeking social support, learning a form of meditation that provides deep rest and in the process a calmer and wider perspective, learning to practice some form of mindfulness such as simply watching thoughts and feelings without engagement, or spending quiet time outdoors. Therapists can reduce shame by normalizing the intensity of feelings in those like themselves who were born highly sensitive. This in turn can provide hope that if others with the trait have found ways to handle these feelings, they can too.

Learning to prevent some of the causes of intense emotions is also helpful. They may be due ultimately to overstimulation, which can lead to exhaustion, anxiety, and depression. For example, clients may plan to move in order in order to take a new job, and even may plan a wedding to precede the move. They may attend a university far from where they grew up, or plan an extended trip with friends on very little money. The major transitions involved, including simply moving, will certainly provoke intense emotions that must be planned for.

Emotions can also be triggered by emotional schemas ("complexes") around earlier distressing events. "We fear what has already happened." By understanding these triggers and accepting, sometimes over and over, that the events were unavoidable, the resulting schemas will generate less emotion with time.

Of all of these methods, normalizing the intense feelings by the therapist may be especially important, particularly if it comes as an expression of kindness and caring. Even if the emotions are due to personal history, as they usually are, differential susceptibility explains it as normal and something the client cannot be blamed for, even if he or she is the only one who can moderate it.

3.2.1 Sensitivity to criticism

If HSPs follow an innate strategy of careful observation in order to avoid mistakes, of "doing it once and doing it right," then it would follow that they would be more attentive by feedback, whether by simply seeing the outcome or by receiving feedback from others. Feedback is the essence of trial-and-error learning, aiming towards no errors. To be oriented to feedback in this way, one would need a strong motivation to get it right and not do it wrong. Strong motivations lead to strong emotions about outcomes.

To see the impact of feedback in an experiment, we (Aron et al., 2005) randomly assigned students to receive positive or negative feedback about their achievement on an "aptitude test," after which they filled out a mood checklist (designed to seem unrelated to the test), and were then debriefed. Students scoring high on the HSP Scale had stronger emotional reactions to both kinds of feedback than the nonsensitive students, who had almost no reaction to either.

The strong reaction to criticism affects therapy as well. The work requires accepting a paradox, that they are truly good, valuable, deserving of love, and that they need help with a serious problem. As therapists we work on a continuum, at one end providing support and reassurance and at the other end discussing problems in depth. With sensitive clients, one can feel stuck at the support end because the client hears anything else as devastating criticism, feeling that they are deeply flawed. Thus, working on their reaction to criticism is central to move their therapy beyond being strictly supportive.

The client needs to learn strategies for adapting one's behavior after receiving negative feedback, and even after having had a strong emotional reaction to it. At first, the instinct may be automatically defensive or overcompensating, or far exceeding expectations in order to avoid the slightest error. Often in therapy it may appear that the client is trying to be "good," finetuning their behavior to the subtle cues they have picked up from you. The result is that both of you are happy and happy with each other. But is their growth?

Meanwhile, they may be actively avoiding situations where criticism or hurtful comments might occur, such as engaging in any work in which they might grow through learning, or avoiding social contact with everyone except the most supportive, familiar others or those clearly inferior to them in some important way. Others wanting or needing to be around them learn to avoid expressing irritation or giving them feedback or criticism. This often leaves clients in a false and confusing world where they are not being given the feedback they need, or they may be receiving mixed messages, since they are sensitive enough to know something is wrong. People seem to like them, yet never call them back; supervisors say their work is fine, but they are not promoted.

Therapists are hampered by not knowing what actually occurred in an interaction outside of therapy. Perhaps the criticism or comment was baseless and the therapist wishes to support the client. Or the criticism might be one the therapist would agree with. Or the feedback was objectively correct: The client lacked the knowledge or skilled required. But the feedback may have been delivered too harshly for a sensitive person. It is especially difficult to know while the emotions are still raw, so first basic self-esteem must be restored, avoiding stating your opinion about the criticism's validity. That means providing solace, demonstrating that you value them for who they are, enough to care how they feel.

With my own clients, when they are angry, sometimes I relate what I have learned over the years from the experience of receiving editing comments that have upset me, even enraged me, I always wait 24 hours before deciding the validity of the suggestions, and often they are at least partly valid. The wait gives me time to cool down and make use of what I agree with. This works in personal relationships as well.

The goal is to work together on thinking about the validity of a comment or criticism. They need to be able to see when the source was a good one, when it was distorting the situation but partly right, or was just plain wrong. There is always a message in any feedback, but it may be that the client needs to learn to ignore the comments of some people. People are noticing and expressing themselves for any number of reasons. What was the motivation behind the comment? Were they trying to be helpful or was it more about their own needs or state of mind? Was the person just tired and stressed that day? Does this person have particular reason to be jealous, offended, or defensive? (Sensitive clients often cannot imagine being a threat to others.) Does the person criticize almost everyone?

The content of the feedback has to be considered. Was it really negative or did the client just hear it that way? Was it helpful, and was it valid in the sense that the person delivering it was in a position to know? Was it based on a simple misunderstanding? Was this really about the client's sensitivity, and if so, what might be done? When and how to talk about one's sensitivity is discussed now.

The client must do most of the deciding here, with the therapist asking the questions that the client needs to consider while resisting drawing conclusions for the client. If the client thinks the comment was invalid, but it is repeated at other times or by other people, they will re-evaluate it on their own. In all of this process of exploring and then reframing, it is the therapist's respect for the client that is mostly likely to lead eventually to less shame and defensiveness, and higher self-esteem. There will be no need to reframe the therapy experience. It will be the frame for a more objective understanding of other experiences.

4 Personal relationships, career, and the workplace

These two topics are considered together because many of the issues overlap. Indeed, they require putting into practice how to avoid overstimulation, set boundaries, regulate emotions, and maintain self-esteem. What is added is dealing with those who are not so sensitive, an issue for both areas of life. Indeed, over the past hundred years these two areas of life, romantic relationships and careers, have changed in remarkably similar ways: the individual is freer to choose among more choices and the resulting choice should bring meaning and satisfaction, not just convenience, financial gain, or increased influence of the extended family. Highly sensitive people must think longer about these choices, and may make mistakes before finding the right person or career. After that, developing relational skills that make their sensitivity an advantage, rather than a liability, becomes crucial.

Relationships and work are also not treated in greater depth here because they are each discussed at length in popular books for the highly sensitive, such as *The Highly Sensitive Person in Love* as well as the book for therapists, *Psychotherapy and the Highly Sensitive Person*.

First, it is worth considering the strengths that highly sensitive people can bring to these areas. In a study of parenting (Aron, Aron, Nardone, & Zhou, 2019) while we found that highly sensitive individuals found parenting more difficult, they also saw themselves as more attuned to their children and more creative with them. Both at home and at work, greater empathy, as part of their stronger emotional responses, allows for a better understanding and rapport with others. Creativity is a natural result of processing things more deeply, until new perspectives and solutions are found. Insights from deep processing can also make HSPs skilled at identifying and solving relationship problems. The various brain studies indicate an attention to details, as well as a sense of the overall patterns of stimulation. Sensitive clients who have established good boundaries and have an understanding of managing relationships with those who are not highly sensitive will be able to bring these strengths to the world with good results.

4.1 Research on relationships

Assuming that those high in SPS process information more deeply and therefore might prefer to do so in relationship conversations, Aron, Aron, Jagiellowicz, and Tomlinson (2010) explored whether SPS would be associated with boredom in relationships, specifically when there is a lack of deep conversations. The results were stronger for women, but significant for both genders. Interestingly, while boredom typically has a negative effect on their relationships, there was no association between their boredom and their marital satisfaction.

In 1998 I conducted a survey of "Temperament and Sexuality" for a book on SPS and relationships (Aron, 2000), where the extensive findings are well summarized. The survey was conducted using a SPS newsletter mailing list of approximately 600 sensitive persons (approximately 450 responded). They were asked to complete the survey themselves, and also, in order to obtain a control group, they were asked to give the same questionnaire to someone else, not highly sensitive, to be returned separately.

The HSP Scale scores were significantly related, for example, to feeling that sex has a sense of mystery or power; finding it difficult to return abruptly to ordinary activities after sex; not being aroused by strong, explicit sexual cues. On the other hand, HSP Scale scores were not related to, for example, number of sexual partners lived with; duration of most recent sexual event; sex being one of the most satisfying parts of life; frequency of orgasm and masturbation; or having been sexually abused, and if abused, having these experiences affect your life; plus there were no differences in reporting a sexual dysfunction (e.g., such as lack of interest, impotence, premature ejaculation) or feelings of satisfaction, worry, excitement, or guilt during sex.

Interestingly, highly sensitive women differed from those without the trait in that sex was less likely to be associated with feeling sad, scared, or afraid and more likely to be associated with feeling loved; and they considered more the impact of a sexual relationship on the other person, were more cautious about sexually transmitted diseases or pregnancy, needed to feel loving toward a partner in order to enjoy sex, and were less able to take sex lightly. Perhaps the lack of fear and sadness regarding sex was due to their having processed deeply the impact of a sexual encounter before engaging in it.

As with the parenting research, we can imagine that some of these results apply to work as well. The highly sensitive, seeming to enjoy deep meaning in whatever they encounter, are probably easily bored when not in jobs that use the full range of their abilities. In unpublished data, they differed in the extent to which work was meaningful, and perhaps their caution in sexual relationships also applies to caution in the workplace about those with whom they share ideas or information (and probably sexual relationships with colleagues as well).

Of course, most of the persons in these studies were not clients in psychotherapy, but still this information may be useful in understanding their relationships.

4.2 Adjusting to those who are not as sensitive

While there are certain problems in a dyad of two highly sensitive people in love, the focus here will be on relationships between the highly sensitive and those who are less sensitive. A fundamental difference is in volume. Earlier, we discussed the therapist's volume—not only loudness but brusqueness, directness, curtness. The highly sensitive have lower volume, often giving hints or suggestions, and can be surprised when someone less sensitive does not respond as they expected, as well as being distressed by high volume in others. In the inevitable conflicts at home or in the workplace, the highly sensitive must learn to adjust their volume, remaining respectful and polite, but also being firm and clear, raising their volume until they are heard.

If the volume of either person becomes high, arousal will be high as well. It is generally recommended that the pair take a time out until a pulse rate over 100 for either person is an objective indicator and reducing this takes about 20 min (Fishbane, 2011). This is extremely important for highly sensitive partners, and should be agreed upon at a time when they are not in conflict. Returning to the discussion is part of this agreement, so that neither will feel abandoned or walked out on.

In the work place, a sensitive employee can sometimes ask for a time out. "I think this discussion is going to be more productive after we take a pause," promising to return to the discussion in a half hour or whenever time it can be arranged.

Time outs are especially important because often the less sensitive person in a long-term relationship has learned, usually unconsciously, that by escalating the conflict and increasing his or her volume, the sensitive partner will back down. His or her reasoning will be less effective, which is then taken advantage of still further. "What you are saying makes no sense." When the sensitive person has had a chance during a timeout to develop better responses to bring to bear in the discussion, they are more likely to negotiate a resolution that includes meeting more of their needs.

All of this brings up the subject of power, or influence. For some of the reasons already discussed, sensitive clients often assume a low rank in any group hierarchy, whether it is two or twenty. Changing this is a slow process of raising self-esteem, teaching about group processes and alliances, examining past experiences that have generalized and now stand in the way, and helping the client see that he or she has much to offer the group, so that it is almost a moral obligation to gain more influence. Rising in rank can be unobtrusive and spontaneous as the client learns to avoid overstimulation and to set boundaries by saying "no." Those higher in rank have the right to more boundaries. When clients confidently assert their boundaries (in appropriate situations, where the client already has implicit power, and makes it sound fair and reasonable), they automatically gain respect and influence.

The client does need to be prepared for strong resistance, however. Those in power, including when it is a relationship partner, will not want a change in the hierarchy, since it means they lose influence. They will certainly resist discussing a specific change that is to their disadvantage, such as helping more with the children at bedtime when the sensitive parent is overwhelmed and no longer effective. But once the client is aware of power issues (which are in every relationship, even a loving one), they will gain skill in seeing what is happening and insisting that fairness prevail.

Given these power issues, clients should not assume that partners will be delighted by the client's discovery of his or her trait. It can imply a distance between them, a change in how they spend time together, and above all, a reason for the sensitive partner to have his or her needs respected and some of those needs will surely be inconvenient for the other. Hence, bringing up one's sensitivity even in an intimate relationship should be approached with caution, and when it is not met with enthusiasm, the client needs to understand why, as a lesson in interacting with those who are less sensitive. The sensitive client must also learn not to use the trait as a sort of club, or as an excuse to avoid anything that is not pleasant for them.

At work, the same potential for shifts in power exist, even when bringing the subject up with a supportive supervisor. The client must learn to think of who, when, and how to discuss the trait, and possibly not to do so at all. Parts can be mentioned without the whole, and pairing a need with an advantage to the other is especially helpful. "I hear that you are pleased with my creativity [productivity; skill with clients, etc.], and that's why I think I'd like to work more from home—I am sure that with quiet and the lack of a commute my work could be even more creative [productive, etc.]." Certainly, requests for special consideration should only be made when clients are confident of being highly valued in the organization, something they often do not realize until it is examined in therapy.

Role-playing responses to issues arising around their sensitivity can also be helpful. For example, if called shy or unsociable at a gathering, they might say, "I don't think of myself that way so much—to me it is more like wanting to observe before entering in. And I am certainly social in that I do thoroughly enjoy a good one-on-one conversation, probably more than situations like this where I don't know anyone yet, and even if I wanted to know them better, it would be difficult to have a deep conversation."

If told they are overly emotional, they can say, "I appreciate your concern, but most of the time I am glad to have my strong feelings. Many of them are good feelings in situations that maybe others would miss. So, it balances out."

Before a medical procedure: "I'm sure you know this—people vary widely in how sensitive they are to pain. I'm one of those with a very low pain threshold."

When being observed or performing and their nervousness has become obvious: "It usually takes me a little while to get used to being watched. I'll be fine soon."

During a lesson or training: "I'm a little overstimulated by the situation, but I will calm down soon. You'll see." Or "I know myself—I'm learning even if it isn't showing yet." Or, if the situation allows it, "Even more than other people, I respond better to praise."

4.3 Getting off to the right start

Finding the right partner (romantic or vocational) and beginning the relationship is often where highly sensitive clients find difficulty. They may choose what others advise because they are seen as so smart or likeable, but often the advisor has not taken into account their advisee's sensitivity. I have seen many highly sensitive young people who were advised to enter law or medicine because they were excellent students, and perhaps parents worried about their ability to support themselves in an aggressive business world. And most of these I have met found the professions inappropriate, often because they disliked the volume or ethics of others, and the resulting stress. Family or friends may also encourage a romantic relationship for similar reasons, not considering the client's sensitivity. As a result, clients may have a history of changing jobs or careers several times, and a similar history with relationships. Hence, such a history of change among the highly sensitive does not always signal instability as it might in other clients.

Because highly sensitive people tend to make decisions slowly, reflecting deeply, a decision that is life changing, such as selecting or changing careers or making a commitment to marriage, will take time. However, jobs often require a quicker choice, resulting sometimes in more job and career changes. Without rushing them, you can point out that a decision delayed can become a decision inadvertently made. For example, if a client passes the age when one normally is accepted for professional training, or allows a good choice for a partner to slip away. The therapist helps them learn to tolerate risk and uncertainty, all part of life. This includes encouraging them to think of the worst thing that could happen if they made a mistake, which is often not as bad as they imagine, and how they could rectify the mistake, not as impossible as it may seem. (I have even used the analogy of choosing the "river of life," not knowing what dangers may await downstream, or stagnating in the "swamp of indecision.") I have written elsewhere (Aron, 2000) about the many reasons sensitive people struggle with commitment to a relationship. But with career or romance, one can point out that they will learn from every mistake, and perhaps wisdom is the natural life goal for the HSP, not simply happiness or success as measured by others.

In seeking the right match, the therapist will need to help them find the right level of openness for encouraging mutual self-disclosure and the right topics. They may have deep thoughts that they express too quickly, or more often, hold back what might be personal or authentic that might intrigue the other (whether on a date or an interview). In particular, since potential partners, and probably employers as well, are often first attracted when they find out the other likes them, the clients may need to disclose some of their deeper positive feelings for the other, even if it a bit ahead of their sense of being able to commit.

Once the relationship is established, volume, adequate downtime, and conflict are the main issues, and we have already discussed these. Downtime (whether 5 min, an hour, or a day) should always be taken in ways that does not leave the partner (or employer) feeling abandoned in general (or in the middle of a shared task). This client can say something genuine about appreciating the relationship (job/coworker/team), along with being clear about when they will return and how pleased they will be to return feeling refreshed.

It may be necessary for the client to learn to speak up about ways he or she is contributing to the relationship or the work place, especially in the particular ways a sensitive person does usually contribute most. This should be done using the lowest, most modest volume that receives a sincere response. Enhancing one's influence may avert later conflicts that are implicitly about who should have more influence.

Another issue is that clients may notice flaws in the other(s) or have better ideas than those being expressed, and have to choose between speaking up and being accused of being overly critical or disregarding the ideas of others, or holding back and feel decreased respect for the others or simply less involved and committed. This also holds true for small habits others may have that bother the more sensitive person (e.g., jingling change in a pocket, chewing with mouth open, glancing at cell phone while the sensitive person is speaking). The less sensitive person may never have such complaints of the more sensitive. The clients learn through discussion of several situations with the therapist about their choice of speaking up or not, either in the past or which to choose in the future. Usually this comes down to choosing one's battles, and maintaining respect for others and possibly not injuring their feelings if they deserve respect for other reasons, even if they do not always see the subtleties that your highly sensitive client does.

There are also work places and relationships in which power issues predominate as a culture, with no remorse. Since aggressively competitive environments do not generally suit the highly sensitive, in this case clients may find the relationship with a partner or one or more business associates so hostile, even bullying, that they have to leave. If this happens several times with jobs, your client may want to explore the possibility of self-employment. If it happens in repeated relationships, there is always the possibility of choosing to remain single and still having close friends, or living separately from a long-term romantic partner.

Hopefully this blending of career/workplace and close relationships issues will help therapists gain a bigger picture of the issues their highly sensitive clients face in both domains, and to sort out personal generalized problematic thoughts and behaviors from temperament issues.

5 Conclusion

Working with highly sensitive clients offers numerous rewards. Merely providing information about their trait that they did not have quickly enhances the client's self-esteem and the therapeutic alliance. Further, success is likely to mirror the research on their being especially benefited by interventions in general. They tend to enjoy therapy, even when it brings painful times, because this may be the only place where the client experiences genuinely deep conversations. These clients are generally respectful of the therapy frame and the therapist's boundaries and personal preferences.

At the same time, they provide special difficulties, regardless of how disturbed they may be by personal stressors and trauma. They may be reluctant to provide useful feedback about the therapeutic relationship or therapy progress, even when asked. They may be repeatedly overstimulated by exploring what must be explored.

Above all, many of their problems, especially men's, are created by the attitudes of others about sensitivity, now and in their past. This is due to ignorance of the trait and of temperament differences in general that will take time to change on a global scale. Sensitive clients however may find some comfort in knowing that some of the struggles they face in life are due to others' ignorance and lack of understanding about sensitivity. Nevertheless, it is especially difficult and damaging to the self-esteem of highly sensitive individuals when hurtful comments or rejections occur. Therefore, it is important to increase awareness of the trait and provide the highly sensitive with adaptive strategies for thriving in a world that is only slowly realizing the reality of this trait and the many benefits it has to offer.

References

Acevedo, B., Aron, E., Pospos, S., & Jessens, D. (2017). The functional highly sensitive brain: A review of the brain circuits underlying sensory processing sensitivity and seemingly related disorders. *Philosophical Transactions B*, https://doi.org/10.1098/rstb.2017-0161.

Acevedo, B. P., Aron, A., Fisher, H. E., & Brown, L. L. (2012). Neural correlates of long-term intense romantic love. *Social Cognitive and Affective Neuroscience*, 7(2), 145–159.

Aron, E. N., & Aron, A. (1997). Sensory-processing sensitivity and its relation to introversion and emotionality. *Journal of Personality and Social Psychology*, *73*(2), 345.

Aron, A., Fisher, H., Mashek, D. J., Strong, G., Li, H., & Brown, L. L. (2005). Reward, motivation, and emotion systems associated with early-stage intense romantic love. *Journal of Neurophysiology*, *9*(1), 327–337. https://doi.org/10.1152/jn.00838.2004.

Aron, A., Ketay, S., Hedden, T., Aron, E. N., Rose, M. H., & Gabrieli, J. D. E. (2010). Temperament trait of sensory processing sensitivity moderates cultural differences in neural response. *Social Cognitive and Affective Neuroscience*, *5*(2–3), 219–226. https://doi.org/10.1093/scan/nsq028.

Aron, E. (2000). *The highly sensitive person in love*. New York: Broadway Books.

Aron, E. N., Aron, A., Jagiellowicz, J., & Tomlinson, J. (2010). *Sensory processing sensitivity is associated with boredom in close relationships*. In: *Paper presented at the International Association for Relationship Research Conference, Herzliya, Israel.*

Aron, E. N., Aron, A., Nardone, N., & Zhou, S. (2019). Sensory processing sensitivity and the subjective experience of parenting: An exploratory study. *Family Relations*, *68*(4), 420–435.

Bakker, K., & Moulding, R. (2012). Sensory-processing sensitivity, dispositional mindfulness and negative psychological symptoms. *Personality and Individual Differences*, *53*(3), 341–346.

Baumeister, F. R., Vohs, D. K., DeWall, N. C., & Zhang, L. (2007). How emotion shapes behavior: Feedback, anticipation, and reflection, rather than direct causation. *Personality and Social Psychology Review*, *11*, 167–203.

Belsky, J., & Pluess, M. (2009). Beyond diathesis stress: Differential susceptibility to environmental influences. *Psychological Bulletin, 135*(6), 885–908. https://doi.org/10.1037/a0017376.

Brindle, K., Moulding, R., Bakker, K., & Nedeljkovic, M. (2015). Is the relationship between sensory-processing sensitivity and negative affect mediated by emotional regulation? *Australian Journal of Psychology, 67*(4).

Chen, X., Rubin, K. H., & Sun, Y. (1992). Social reputation and peer relationships in Chinese and Canadian children: A cross-cultural study. *Child Development, 63*, 1336–1343.

Chen, X., Cen, G., Li, D., & He, Y. (2005). Social functioning and adjustment in Chinese children: The imprint of historical time. *Child Development, 76*(1), 182–195.

Fishbane, M. D. (2011). Facilitating relational empowerment in couple therapy. *Family Process, 50*(3), 337–352.

Greven, C., Booth, C., Lionetti, F., Acevedo, B., Aron, E., Pluess, M., … Homberg, J. (2019). Sensory processing sensitivity in the context of environmental sensitivity: A critical review and development of research agenda. *Neuroscience and Biobehavioral Reviews, 98*, 287–305.

Hillesum, E. (1983). *An interrupted life: The diaries of Etty Hillesum* (Trans. Arnold J. Pomerans). New York: Pantheon.

Larsen, R. J., & Prizmic, Z. (2004). Affect regulation. In R. F. Baumeister & K. D. Vohs (Eds.), *Handbook of self-regulation: Research, theory, and applications* (pp. 40–61). New York, NY, US: The Guilford Press.

Nocentini, A., Menesini, E., & Pluess, M. (2018). The personality trait of environmental sensitivity predicts children's positive response to school-based antibullying intervention. *Clinical Psychological Science, 6*(6), 848–859.

Pluess, M., Assary, E., Lionetti, F., Lester, K. J., Krapohl, E., Aron, E. N., & Aron, A. (2018). Environmental sensitivity in children: Development of the Highly Sensitive Child Scale and identification of sensitivity groups. *Developmental Psychology, 54*(1), 51.

Pluess, M., & Belsky, J. (2013). Vantage sensitivity: Individual differences in response to positive experiences. *Psychological Bulletin, 139*(4), 901–916. https://doi.org/10.1037/a0030196.

Pluess, M., & Boniwell, I. (2015). Sensory-processing sensitivity predicts treatment response to a school-based depression prevention program: Evidence of vantage sensitivity. *Personality and Individual Differences, 82*, 40–45. https://doi.org/10.1016/j.paid.2015.03.011.

Zeff, T. (2010). *The strong sensitive boy: Help your son become a happy, confident man*. Prana Publishing.

CHAPTER 7

Clinical characteristics of misophonia and its relation to sensory processing sensitivity: A critical analysis

Dean McKay[a], Bianca P. Acevedo[b]
[a]Department of Psychology, Fordham University, Bronx, NY, United States
[b]Neuroscience Research Institute, University of California, Santa Barbara, CA, United States

Contents

1 Basic conceptualization of misophonia 166
2 Defining the boundaries of misophonia 167
2.1 Assessment of misophonia 170
3 Emotional reactions associated with misophonia—Disgust and anger 173
4 Association of misophonia with other disorders 174
5 Sensory processing sensitivity and misophonia 179
6 Interventions and treatment of misophonia 180
7 Conclusions and future directions 181
7.1 Couching misophonia in fundable terms 182
7.2 Getting the word out 182
References 183

Misophonia is a serious and disabling condition marked by intense aversive emotional reactions to certain sounds (such as respiratory noises, chewing, and other human-caused noises). Recently described by Jastreboff and Jastreboff (2002), misophonia can be literally translated to "hatred of sounds." However, that is really an overstatement, as individuals suffering from misophonia report strong aversion to specific sounds, as well as emotional reactions resulting from the anticipation that these sounds might occur (such as food being served, preparing for bed, sitting at a computer). Shortly after the condition was first characterized in the scientific literature, a flurry of case treatment studies appeared, with highly varied interventions. There has also been a developing programmatic approach to understanding the nature of misophonia, common co-occurring features, methods of assessment, and an emerging systematic protocol-driven set of procedures for treatment.

The Highly Sensitive Brain
https://doi.org/10.1016/B978-0-12-818251-2.00007-2

Given the relatively recent characterization of misophonia in the scientific literature, this chapter has the following broad aims. First, to define the boundaries of misophonia, compared to other auditory problems, using a psychological context. Second, to describe the research on commonly co-occurring psychological conditions, as well as potentially complicating conditions. And third, to suggest specific future directions for research that will illuminate best methods for clinical intervention with misophonia.

1 Basic conceptualization of misophonia

A fundamental feature of misophonia involves adverse reactions to specific sounds, and not to all sounds. Given the additional central problem misophonia sufferers report, namely, anticipatory reactions to situations (or triggers, in the common language parlance of sufferers), the full etiology of the condition may be conceptualized through established learning theory principles.

At a fundamental level, several models have been suggested for misophonia. One theory suggests that misophonia could be construed as originating from two factors (Mowrer, 1960). In this model, the aversive reaction to a specific sound is acquired through classical conditioning. Specifically, a target sound (unconditioned stimulus, UCS) leads to an aversive response (unconditioned response, UCR). When paired with other signals (i.e., environmental stimuli associated with the sound), then these stimuli serve as conditioned stimuli (CSs). These CS could act then as "triggers" for the aversive response (now a conditioned response, CR). Following this, the avoidance and other associations, such as family and friends associated with the trigger, would serve to maintain the aversive responses through negative reinforcement (Skinner, 1938).

The two-factor model is not a complete depiction of misophonia, however, since it does not account for individual differences whereby some individuals may develop strong aversions to certain sounds, while others may merely find these sounds irritating, but generally tolerable. One perspective that can resolve this issue involves stress tolerance. Specifically, it has been shown that there is wide variability across individuals in stress reactivity, with stress broadly defined as a reaction to threat. When stress is sustained over a long period, the degree that one is physically affected by it varies based on coping strategies (Lazarus & Folkman, 1984). Included in this model is the extent that one responds emotionally to stress, which in turn can determine the extent that one evaluates the stress experience as aversive. These responses have implications for physiological responses (Gorka, LaBar, & Hariri, 2016) and over a long term, can result in deterioration in physical health (reviewed in Fink, 2009).

The described conditioning perspectives have implications for specific neural regions. Accordingly, a neuroscience-informed conceptualization of misophonia that integrates learning experiences can be advanced, and with additional research a potential circuit model may effectively describe the condition. Specifically, given that auditory processing involves the limbic system, threat appraisal is a prominent component of misophonia. The limbic system and associated neural regions (such as the amygdala, parahippocampus, and insula) have been implicated in several other auditory conditions, such as phonophobia, and as a result have been hypothesized as being related to misophonia (Palumbo, Alsalman, de Ridder, Song, & Vanneste, 2018).

2 Defining the boundaries of misophonia

As of this writing, misophonia does not exist as a formal diagnosis in any taxonomic system. As a result, researchers are still working out the basic features of this condition, including the possibility that it exists as part of some other syndrome or disorder (Taylor, 2017). Schröder, Vulink, and Denys (2013) proposed diagnostic criteria for misophonia. These criteria cover six broad areas[a]:

- Presence or anticipation of specific sounds that leads to an aversive physical reaction that includes irritation and/or disgust, leading to anger.
- The anger leads to loss of control, with potentially aggressive outbursts.
- The individual recognizes that the anger and/or disgust is excessive or more than warranted by the sound.
- Avoidance of sounds, and associated situations, is typical; if sounds and/or situations are endured, it is with intense discomfort.
- The emotional reactions lead to significant distress and/or interference in daily life.
- The emotional reactions are not better accounted for by another disorder.

Schröder et al. (2013) provide a detailed accounting for differentially assessing misophonia from other conditions that may resemble it, which include: specific phobia, including phonophobia (fear of sounds);

[a]The categories listed here were adapted from the proposed criteria by Schröder et al. (2013). See original article for exact text of their proposed criteria.

posttraumatic stress disorder; social phobia; obsessive-compulsive disorder; intermittent explosive disorder; personality disorders characterized by aggression (such as borderline and antisocial); and autism spectrum disorders (ASDs) and other sensory processing disorders.

While it is important that specific taxonic features of misophonia have been outlined, there remain substantial questions regarding the basic nature of this problem, such as whether it is better accounted for within the context of other disorders. To this end, misophonia may be at-risk for contributing to the criticism that diagnostic systems, such as the Diagnostic and Statistical Manual of Mental Disorders (DSM-5; American Psychiatric Association, 2013) or International Classification of Diseases (ICD-10; World Health Organization, 2015), pathologize minor personal idiosyncrasies rather than identify genuine syndromes (Stein, Phillips, Bolton, & Fulford, 2010; Taylor, 2017; Taylor, Asmundson, Abramowitz, & McKay, 2009). This issue is not unique to misophonia.

Mental health practitioners have long recognized the limited syndromal validity of the existing diagnoses. Unlike other classification systems, the psychiatric nomenclature is developed by committees, rather than by theory-driven signs and symptoms. Further, research has shown that the committees charged with crafting the diagnostic criteria have inherent conflicts of interest with the pharmaceutical industry, thus leading to potential bias in favor of overpathologizing ordinary variations in human experiences and behavior (Pilecki, Clegg, & McKay, 2011). Research employing taxometric analyses—which identify the extent that an indicator of psychopathology exists on a continuum, and thus represents a range of severity from normal to pathological (but not necessarily syndromal) or as a discrete entity, thus allow for classification of a putative disease entity (see Waller & Meehl, 1998)—has revealed few discrete psychopathological entities. Further, the current descriptive taxonomy (after which the proposed misophonia criteria are modeled) are atheoretical. Therefore, the existing models of diagnosis do not provide guidance for mechanisms[b] of psychopathology nor interventions.

In reaction to the failure of the DSM and ICD systems to validly define psychopathology, the National Institute of Mental Health (NIMH)

[b]Although it is beyond the scope of this chapter, mechanisms of psychopathology and treatment have been increasingly recognized as essential in determining evidence-based treatment. Kazdin (2007) provided a detailed understanding of the methodological and programmatic steps involved in identifying any putative mechanisms of psychopathology and treatment.

launched the Research Domain Criteria (RDoC) initiative which aims to define psychiatric conditions based on valid indicators of psychopathology (i.e., neuroticism) and evaluate these indicators in a matrix of categories (i.e., negative and positive valence systems, cognitive systems) and in a hierarchy of assessment domains (i.e., genes, cells, molecules, neural circuits, etc.) (Insel et al., 2010). Fig. 1 shows the basic structure of the RDoC matrix. This system emphasizes a dimensional perspective on psychiatric conditions, but also stresses that any condition should be distinguishable in unique ways along multiple levels of assessment (see Cuthbert & Kozak, 2013; but see also Lilienfeld, 2014). This means that a comprehensive research program would necessarily focus on multiple levels of assessment, and further categorize these into multiple construct assessment domains.

In addition to the RDoC initiative, research over the past two decades has revealed that psychopathology can be categorized into a few broad dimensions, and that these large dimensions can be further delineated in a hierarchical arrangement (Krueger, Caspi, Moffitt, & Silva, 1998). This refined model, referred to as the Hierarchical Taxonomy of Psychopathology (HiTOP; Conway et al., 2019) sets the occasion for identification of specific valid indicators of psychopathology, which therefore can be integrated with the aforementioned RDoC initiative. HiTOP arranges all indicators of psychopathology under two broad factors; internalizing and externalizing, with internalizing comprising the two subfactors of fear and anxious misery. Where misophonia might fit in this model is not known. Given that the condition is characterized by intense anxiety and disgust, it could easily fit under anxious misery. On the other hand, based on the proposed criteria by Schröder et al. (2013), where anger and potential aggression are noted, it could also be considered an externalizing

Construct domain	Genes	Molecules	Cells	Circuits	Physiology	Behavior	Self-report
Negative valence systems							
Positive valence systems							
Cognitive systems							
Systems for social processes							
Arousal/modulatory systems							

The full matrix, which has additional subcategories for each construct domain, can be accessed at https://www.nimh.nih.gov/research-priorities/rdoc/constructs/rdoc-matrix.shtml

Fig. 1 Research domain criteria matrix.

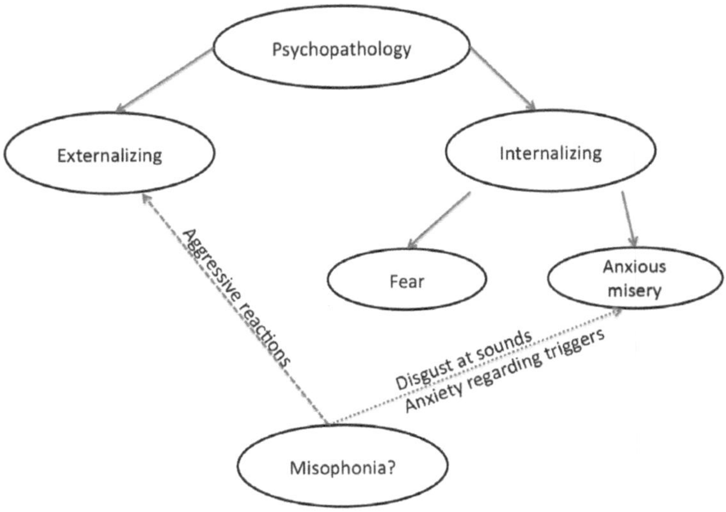

Fig. 2 Hierarchical taxonomy of psychopathology (HiTOP) with hypothetical placement of misophonia.

problem. Fig. 2 schematically lays out the highest order factors in HiTOP, with hypothetical placement of misophonia. However, at the most fundamental level, this requires valid assessments of misophonia.

2.1 Assessment of misophonia

At present, there are two self-report measures of misophonia. While neither has extensive psychometric support as yet, the two measures have been used in research investigating misophonia. There is also a symptom severity checklist in development.

The Amsterdam Misophonia Scale (*A-MISO-S*; Schröder et al., 2013): The A-MISO-S is a checklist that assesses severity of symptoms along the dimensions of the proposed criteria for misophonia, such as sounds that trigger emotional reactions and interference in functioning as a result of different sounds. There are six questions, each rated from 0 (none) to 4 (Extreme). As yet, there are no psychometric data on the A-MISO-S. In a group cognitive-behavior therapy trial, there was significant change in the A-MISO-S following seven sessions (Schröder, Vulink, van Loon, & Denys, 2017). A total of $N = 90$ individuals ($N = 65$ women and 25 men) who sought treatment specifically for misophonia participated. Treatment components were task concentration, which was to focus on other activities when potential aversive sounds may be generated; counterconditioning, whereby positive images or sounds were paired with the target aversive noises; sound manipulation, whereby individuals practice regulating the sounds through an in vivo online task; and relaxation training. Sensitivity to treatment is an important marker of the validity of the measure. However, additional psychometric evaluation of the scale is warranted. It should be noted that since the scale is an assessment of severity along dimensions of the proposed misophonia diagnostic criteria, it is not clear that the scale can function as a diagnostic tool as it does not provide a means for evaluators to rule out other conditions (such as those listed by Schröder et al., 2013). Accordingly, this scale should only be used in conjunction with other interview methods to rule out other conditions that may plausibly be associated with misophonia.

The Misophonia Questionnaire (*MQ*; Wu, Lewin, Murphy, & Storch, 2014): The MQ comprises two broad dimensions: one asks participants to self-rate [from 0 (not at all true) to 4 (always true)] their sensitivity to seven specific types of sounds (i.e., eating, repetitive tapping, vocal expressions of consonants or vowels), and the other including ten questions regarding reactions to the sounds rated at a 1 (rarely true) or higher, for the prior seven sound areas. The scale is scored by summing the items. There are no established cutoff scores that distinguish individuals with or without misophonia. There is also a global severity rating, ranging from 1 (minimal) up to 15 (very severe). The scale comprises two broad factors, corresponding to the self-rated target sounds and the self-rated reactions to sounds. In the initial research involving scale development, which comprised a large (N = 483) undergraduate sample, it was also found that scores were associated with a wide range of psychopathology indicators, notably: sensory intolerance, disability, depression, anxiety, stress, and obsessive-compulsive symptoms. In a follow-up investigation, the MQ was found to have similar properties in a large sample of Chinese college students (N = 415). Specifically, Zhou, Wu, and Storch (2017) found that severity scores were associated with impairment in work, school, social situations, and family engagement, as well as global sensory intolerance, obsessive-compulsive symptoms, anxiety, and depression.

In a separate community sample (N = 628), individuals with significant misophonia had lower levels of obsessive-compulsive symptoms, except for concerns around orderliness and harm avoidance (McKay, Kim, Mancusi, Storch, & Spankovich, 2018). This study employed the same assessment of obsessive-compulsive symptoms as the Wu et al. (2014) investigation. The primary difference was the McKay et al. (2018) investigation was with community participants, while the Wu et al study was with undergraduate students.

Accordingly, this relationship deserves additional investigation to determine the extent that obsessive-compulsive symptoms contribute or are related to misophonia. Individuals with elevated misophonia show higher rates of interpreting interoceptive physical sensations as indicative of possible danger. Collectively, the MQ appears to be a reliable and valid indicator of misophonia severity. However, as with the A-MISO-S, the MQ should not be considered a diagnostic tool, but instead should only be used to assess misophonia severity in conjunction with suitable assessment tools to rule out other possible conditions that could also be expressed with specific sound intolerance.

The Misophonia Severity Index (*MSI*; McKay & Storch, 2018 *in development*): The MSI comprises a symptom checklist that is an elaboration on the one that appears in the MQ, categorized into the following "trigger sounds": mouth/throat sounds, breathing/respiratory sounds, vocal sounds, and other sounds. Following the checklist, there are four clinician-administered severity items that assess the following: frequency, distress, avoidance, and emotional reactions. The full scale is provided in Appendix of this book. At this point, the MSI has been part of two treatment trials that are in progress, one with adults, and one with children (ages 7–17) being conducted in the Compulsive, Obsessive, and Anxiety Program lab at Fordham University. No psychometric or outcome data are yet available.

Summary of assessments: At present, there are only three assessment measures for misophonia, and two of which (A-MISO-S; MQ) have psychometric data to suggest either sensitivity to treatment (A-MISO-S) or reliability and construct validity (MQ). At this point, considerably more research is necessary to determine the full psychometric characteristics of these measures. In order to conduct a full assessment of misophonia, and not only determine its presence but also rule out other possible diagnostic conditions that may account for sound sensitivity, administration of major diagnostic interview tools may be necessary. For example, considering the broad range of potential psychiatric conditions listed in Schröder et al. (2013), administration of the structured clinical interview for DSM-5 (First, Williams, Karg, & Spitzer, 2015) should be considered. Alternatively, if it is possible to rule out several possible conditions through a general clinical interview [i.e., sensory processing sensitivity (SPS), intermittent explosive disorder, personality disorders, autistic spectrum and other sensory intolerances], then administration of the Anxiety and Related Disorders Interview Schedule for DSM-5 (ARDIS-5; Brown & Barlow, 2015), which evaluates anxiety disorders, obsessive-compulsive and related disorders (OCRDs), and depression.

3 Emotional reactions associated with misophonia—Disgust and anger

Schröder et al. (2013) emphasized that misophonia is associated with not only anxiety, but also disgust from the sounds, which in turn may lead to anger. Disgust is a complex emotion that serves the general function of protection from ingestion of harmful substances (see McKay, 2017, for a review). Briefly, disgust comprises classes of stimuli that evoke the emotion. These classes are as follows: food, insects and other animals, sex, body products (such as physical waste), body violations (such as observing exposed organs), and death. Further, disgust follows the laws of sympathetic magic, which allows for the transfer of disgust-related properties from the primary disgust-evoking object to otherwise neutral objects (Rozin & Fallon, 1987). Sympathetic magic comprises the Law of Similarity and the Law of Contagion. The Law of Similarity is the disgusting property that an object takes on when it merely resembles a disgusting object (i.e., soup bowls shaped like toilets). The Law of Contagion is the perceived contamination of an otherwise neutral object after it has had prior contact with a disgust-evoking stimulus (i.e., using a thoroughly washing a soup spoon after it has been used to kill a spider). In addition to the primary disgust elicitors noted, recent research has focused on moral disgust (i.e., Rozin, Haidt, & Fincher, 2009). This facet of disgust has a clearer and more direct line to anger than the other aspects of disgust.

Finally, disgust is experienced differentially among individuals (Olatunji & Broman-Fulks, 2007). For example, one individual might have a higher level of disgust propensity, whereby different disgust-eliciting stimuli and situations more readily provoke disgust. On the other hand, another individual might have higher disgust sensitivity, which is the tendency to interpret physical sensations as indicators of disgust (i.e., a bubbly feeling in the abdomen interpreted to suggest that the stimuli in front of the individual is disgusting).

It has been shown that specific sounds can elicit disgust, particularly in relation to food and bodily noises. As disgust emerged to protect from ingestion of potentially harmful substances (discussed in McKay, 2018), chewing sounds and other food consumption noises, which are frequently associated with misophonia, would also prompt disgust responses. In a similar vein, as bodily products represent a significant category of disgust elicitors, chewing and other gustatory noises would be candidates for misophonia reactions given the auditory experience of bodily secretions. If we focus for a moment on chewing per se, the disgust relevance becomes readily evident: chewing involves food consumption, and as one continues to chew, the shift in the nature of the sound is illustrative of the combination of the food with bodily secretions (specifically saliva).

An important question then emerges—"what accounts for anger, if the primary emotion is disgust?" In the proposed criteria for misophonia (Schröder et al., 2013), it is suggested that an initial disgust reaction quickly switches to anger. In this instance, the anger response is directed at the source making the noise. When viewed through this lens, one can see how a misophonia sufferer might show heightened distress around certain individuals (such as family members). If the individual makes the specific target sound that causes distress (namely, disgust) in the misophonia sufferer, they in turn experience anger toward the target individual. This individual may be conditioned to be a trigger for specific sounds, and in some extreme cases there may be avoidance of the individual who makes the distressing noises. Indeed, this can be extremely disruptive to the basic fabric of society, the family structure, and one's close relationships.

4 Association of misophonia with other disorders

The proposed diagnostic criteria for misophonia, and the extant research on the correlates between misophonia severity and other psychopathological states are suggestive of a connection between the condition and other diagnoses or conditions. Much of this research is either single-case or controlled-case histories.

Anxiety and anxiety disorders: Broadly, the class of anxiety disorders refer to a collection of conditions marked by heightened anxiety for specific stimuli (i.e., specific phobia, SP), situations (i.e., social anxiety disorder, SAD), or interoceptive experiences [i.e., worry in generalized anxiety disorder, GAD; or panic attacks in panic disorder with agoraphobia (PDA)].[c] However, in searching the research literature, there were no reported cases or systematic investigations of misophonia in relation to any of the anxiety disorders. Accordingly, what will follow is a brief review of how anxiety, broadly conceived, may be associated with misophonia, followed by a conceptualization of the manner in which other anxiety disorders may be present in conjunction with misophonia.

[c]There has been extensive research on the nature and treatment of anxiety disorders, including conceptualization and treatment (Barlow, 2002) and gender differences (Craske, 2003).

General anxiety: As noted previously, several reports have shown misophonia to be associated with elevated levels of anxiety in general (McKay et al., 2018; Wu et al., 2014; Zhou et al., 2017). Further, one report found heightened anxiety sensitivity (Taylor, 1999) in individuals as a function of severity of misophonia symptoms (McKay et al., 2018). Briefly, anxiety sensitivity is the extent that individuals appraise changes in interoceptive experiences [i.e., change in heart rate (HR), respiration] as potentially harmful. This construct was originally formulated to describe the basic pathophysiology of panic (Reiss, Peterson, Gursky, & McNally, 1986). Since that time, it has been shown to have wide applicability to psychopathology, particularly the full range of anxiety disorders (Olatunji & Wolitzky-Taylor, 2009).

Conceptualization of anxiety disorders and misophonia: In the conditioning model described at the beginning of this chapter, it was highlighted that aversive emotional responses serve as a vehicle for acquiring the conditioned negative reaction to specific sounds. One major aversive reaction is anxiety. As it was also noted that there are individual differences in how people experience anxiety (that is, some find it more aversive than others), the sensitivity to this emotional state can readily set in motion a conditioned aversive reaction to specific sounds. These aversive reactions can resemble the different anxiety disorders.

Specific phobia (SP): Individuals with misophonia report that they have "triggers" that prompt their aversive reactions to sounds. That is, there are different environmental features that signal the possibility that the unpleasant trigger sound is about to occur. For example, if someone finds chewing sounds distressing or bothersome, the sight of plates and a dinner table setting might serve as a trigger for adverse reactions. These could range from uncomfortable feelings, stress, and fleeing of the situation. In this way, individuals with misophonia might avoid certain places, situations, or objects in a manner comparable to those with a specific phobia (SP). For example, in a treatment trial in the COAP lab, a woman in her late 20s avoided babies and situations associated with babies, due to her extreme negative reaction to crying. Anticipation of experiencing crying babies in particular was met with the classic SP reactions—tachycardia, dizziness, tingling in the extremities, along with avoidance of situations where she might encounter babies.

Panic disorder (with or without agoraphobia): Panic disorder is characterized by sudden and intense anxiety (panic attacks) that sufferers say seemingly comes "from out of the blue." Following an attack, sufferers experience anticipatory anxiety for future attacks. For many sufferers, places and situations where an attack previously occurred will be avoided, leading to agoraphobic avoidance (discussed in Barlow, 2002). In the case of misophonia, panic symptoms may be experienced when the individual is confronted with triggers for sounds that are intensely aversive, and the sufferer perceives that there is no means to escape the experience.

Generalized anxiety disorder: Most people experience worry, and in normal worry, it serves a useful function in problem solving. That is, worry leads to arousal that orients the individual toward a means for solving some concern (Davey & Wells, 2006). In contrast, pathological worry is associated with a repetitive cycle of the worry experience, with the sufferer perceiving that there is no solution possible, or that any solution generated will likewise be intolerable. When the worry is pathological in nature, there is a cascade of additional physiological symptoms (prolonged heightened HR, blood pressure, muscle tension), and a litany of interpersonal difficulties that lead to general anxiety disorder (GAD) (Heimberg, Turk, & Menin, 2004). In the case of misophonia, sufferers may develop worries over the next occasion they will confront either a trigger or a sound that will be intolerable, and with little recourse for how to cope with these possible encounters. In order to manage the anxiety that would result from these worries, a misophonia sufferer might develop negative (versus adaptive) coping strategies to arrange situations to avoid possible encounters with triggers and sounds. In turn, the negative coping strategies may further amplify a misophonic's life issues resulting in diminished quality of life, life satisfaction, and life functioning. Although this has not been formally examined in misophonia, for highly sensitive individuals, studies have observed diminished quality of life and close relationships that result from maladaptive regulatory abilities, which are mitigated among high SPS individuals with positive, nurturing, and supportive childhood environments.

Social anxiety disorder: Formerly called social phobia, it was changed to social anxiety disorder (SAD) when it was recognized that the manifestation of social avoidance was a more complex syndrome than what was implied by social phobia. SAD is associated with significant social disability that may result in heightened liability for substance use (i.e., alcohol; Miloyan & Doorn, 2019). The primary aversive experience in SAD is the risk or actual experience of negative evaluation by others and the feeling of embarrassment. In the case of misophonia, a sufferer may manifest SAD through a concern that peers or family members will judge the evident aversive reactions to certain sounds negatively. Also, individuals with misophonia (like SPS) may develop SAD as they may be hypervigilant, tend to ruminate more (due to paying more attention to stressors), or experience fear not being able to escape aversive social situations, as discussed in Chapter 4 of this book.

Conceptualization of obsessive-compulsive and related disorders with misophonia: In the DSM-5, a new category was formed, the obsessive-compulsive and related disorders (OCRDs), which comprised the following diagnoses: OCD, body dysmorphic disorder (BDD), hoarding disorder (HD), Trichotillomania (TM), and excoriation disorder (ED). In the prior editions of the DSM, OCD was classified in the anxiety disorders and HD was a symptom of OCD. BDD was classified with somatoform disorders. TM was classified with the impulse control disorders. ED did not exist, and is therefore a new diagnostic entity.

In consideration of the basic manifestation of misophonia, there is little clear rationale for the condition to be a causative mechanism in many of the OCRDs, with the exception of OCD per se. OCD is a complex and heterogeneous disorder in its own right. Most experts on OCD recognize that it comprised of subtypes, which can be roughly categorized into the following: checking and harming obsessions; contamination fears and washing rituals; and symmetry and ordering (Abramowitz, McKay, & Taylor, 2005).

As noted earlier, research with a community sample suggests that obsessive-compulsive symptoms tend to be lower in individuals with misophonia, with a specific exception, namely, ordering. Also, harm avoidance, which is a cognitive feature of OCD, could be construed as a feature common to anxiety disorders as well. Therefore, the extent that OCD in particular, or cognitive dimensions that are not unique to OCD in general, are heightened in misophonia warrants additional investigation.

Conceptualization of trauma and stress related disorders with misophonia: As with the OCRDs, the DSM-5 formed a new category, the trauma and stress-related disorders. This category comprised posttraumatic stress disorder (PTSD), acute stress disorder (ASD), adjustment disorders (AD), and chronic adjustment disorder (CAD). As of this writing, there are no documented cases or controlled examinations to evaluate the extent that trauma is associated with the onset or maintenance of misophonia. However, across trauma- and stress-related disorders, a conceptualization may be made to justify the potential onset of misophonia.

Consider the central condition of the PTSD category. The DSM-5 emphasizes that onset of the disorder emerges from a Criterion A stressor, whereby the sufferer was exposed to the threat of death, actual or threat of serious injury, or actual or threat of sexual violence. This could include a direct experience or witnessing such a situation in-person. Many sufferers find that even minor or subtle reminders of the prior trauma activate their fear or panic, thus leading to serious disability. Accordingly, individuals with prior trauma often find that specific sounds associated with the event cause intense distress, leading to avoidance, and also have associated triggers.

Conceptualization of attention deficit disorder with or without hyperactivity in misophonia: As with other disorders reviewed, there is as yet no research specifically examining misophonia in relation to attention deficit disorder (ADD) with hyperactivity (ADHD) or without. In considering the means for understanding the possible relation between ADD with misophonia, it is essential to consider the temporal order of the two. Specifically, if an individual is consistently vigilant for possible triggers of specific aversive sounds, this would drain cognitive resources that would be used to allocate attention to other environmental events, pleasure, and creative capabilities. At issue is that baseline level of cognitive load that an individual can tolerate. Cognitive load refers to the amount of mental activities that can be engaged in at any given time. Since the amount of cognitive load anyone can handle is finite, anything that draws on extensive cognitive resources will compromise the ability to attend to additional information (see Sergeant, 2000, for a review and conceptualization). Therefore, vigilance for triggers will draw from the finite cognitive resources of an individual, thus creating the appearance that the misophonia sufferer is also struggling with ADD, PTSD, or other related disorders; and perhaps even result in marked interpersonal issues.

Autism spectrum disorders and misophonia: While autism spectrum disorders are broadly characterized by heightened sensory intolerance (Dawson & Watling, 2000), there is limited research on specific sound intolerance of the type associated with misophonia. In one review, it was suggested that there was no difference in the physiological reactivity in individuals with ASDs compared to non-ASD individuals (Stiegler & Davis, 2010). This suggests that any specific sound intolerances observed in individuals with ASD is likely a consequence of learned associations rather than primary biological processes. Also, with respect to SPS, some research has shown that the two engage distinct neural processes, with the central difference being that SPS is commonly associated with enhanced empathic and emotional processing in response to social events, while ASD commonly shows deficits in processing of social information (for review see Acevedo, Aron, Pospos, & Jessen, 2018).

5 Sensory processing sensitivity and misophonia

As noted, one of the central features of SPS is enhanced empathic processing. High sensitivity is also associated with depth of processing, having a rich inner life, intuition, and being attuned to subtleties in the environment and others. For example, high SPS is associated with more attunement and pleasure in response to aesthetically pleasing art work. However, it may also manifest as discomfort or irritation with loud and chaotic noise, as overlapping with some of the symptoms of misophonia.

In the brain, SPS shows greater activation in response to emotional stimuli in brain regions that coordinate awareness, the integration of sensory information, empathy, and action planning; namely, the insula, inferior frontal gyrus (IFG), middle temporal gyrus (MTG), and premotor area (Acevedo et al., 2014). Also, SPS is associated with enhanced activation of brain regions involved in memory and physiological homeostasis, as well as reward in response to positive stimuli (Acevedo et al., 2014; Acevedo, Jagiellowicz, Aron, Marhenke, & Aron 2017).

In contrast, as described earlier, misophonia is characterized by the strong negative emotional response to specific trigger sounds, which may also come to be associated with other various stimuli, for example, visual and social things, people, or events. However, research suggests that in misophonia there is a differential response, specific to sounds, that may or may not have an auditory basis. That is, where in SPS, heightened sensitivity to sound may be due to the inability to process too much information at any given time, for misophonia the negative response to sound is due to some cognitive representation of the sound trigger. For example, Edelstein, Brang, Rouw, and Ramachandran (2013) showed that in misophonics, skin conductance response (SCR) and aversiveness ratings were increased in response to auditory-only stimuli as compared with controls. Correspondingly, at least one brain imaging study (using functional MRI) measuring HR and GSR along with neural response to trigger sounds, unpleasant sounds (aversive nonmisophonic sounds) among misophonics and control subjects showed that the "trigger" (versus unpleasant and neutral) sounds evoked an abnormal response in individuals with misophonia. Also, in misophonics trigger sounds evoked abnormal functional connectivity between the anterior insula cortex, and a network of regions involved in emotion, memory, and emotion regulation (namely, the amygdala, hippocampus, and ventromedial prefrontal cortex, vmPFC). Trigger sounds elicited heightened HR and galvanic skin response (GSR) in misophonic subjects, which were mediated by AIC activity. Also, brain structural measurements showed greater myelination within the vmPFC in misophonic individuals. Interestingly, no differences were found for unpleasant and neutral sounds in misophonics versus nonmisophonic controls.

In sum, the central characteristics and neural pathways associated with SPS and misophonia are distinct. SPS is characterized by overall greater responsivity to both positive and negative environmental and social stimuli; while misophonia is distinctly associated with aversion to specific trigger sounds. Moreover, SPS in adaptive environments is associated with enhanced empathy and depth of processing, while misophonia may evoke fight/flight responses to trigger sounds associated with disgust, anger, and avoidance. In addition, the relatively few brain imaging studies of SPS and misophonia suggest that some of the neural pathways that mediate response to stimuli are relatively distinct for SPS and misophonia in areas that mediate physiological homeostasis, but overlapping in brain areas associated with emotion, memory, and visceral processing (namely, the amygdala, hippocampus, and AI). However, much remains to be studied in these two emerging fields as it is unknown whether individuals with SPS may also be predisposed to misophonia (as they are with other disorders such as depression and anxiety). To this extent, in the next section we discuss potential interventions for these conditions.

6 Interventions and treatment of misophonia

At present, aside from the aforementioned open group-based treatment trial (Schröder et al., 2017), there are only a series of illustrations of single or small series of case studies. In the treatment trial conducted by Schröder et al. (2017), there were four components of treatment, as described earlier, and resulted in a reduction of primary symptoms.

Another approach that has been recommended, but not systematically investigated with misophonia, is tinnitus retraining therapy (TRT). In the treatment of tinnitus, a condition marked by persistent experiences of buzzing or other persistent steady noises that are not externally generated, TRT has been shown effective (Folmer, Theodoroff, Martin, & Shi, 2014). The treatment comprised several major components. Coping skills are taught to increase tolerance to the sound. Relaxation training is administered to further alleviate distress, along with a broader stress management program, such as improving appraisal of other stressors. In some instances, sufferers are provided ear-worn devices designed to divert attention from the noises.

Somewhat differently, some case study investigations have generally used exposure-based treatments to alleviate misophonia (i.e., Bernstein, Angell, & Dehle, 2013; McGuire, Wu, & Storch, 2015; Reid, Guzick, Germand, & Olsen, 2016). However, anecdotal reports (including those posted on various patient forums) suggest that exposure therapy may actually be a harmful intervention (Palumbo et al., 2018). It is also likely that clinicians administering exposure in uncontrolled contexts do not have the means that would allow for proper pacing or titration for the treatment to prove beneficial. It has been demonstrated that when exposure is administered at a pace where the individual has inadequate opportunities to master the experience, there is often a worsening of symptoms (Foa & Kozak, 1986). However, these case series warrant further systematic evaluation, especially with regard to misophonia. However, basic research on misophonia may also shed light on this as its roots may not be in a learned response that can somehow be unlearned, but rather managed adaptively.

7 Conclusions and future directions

In this chapter, an overview of the psychopathology research on misophonia was covered. Aside from a limited collection of single case treatment illustrations, the research literature is still very much in its infancy in characterizing misophonia. Among the areas yet to be determined: whether it is a distinct diagnostic entity; if it is a unique diagnostic condition, what are its primary defining features; if these defining features can be characterized, what are common comorbid associations; what is the etiology and course of the condition; what are the public health consequences of the condition; and what are the primary mechanisms of the condition that may serve as suitable targets for treatment. This is not necessarily an exhaustive list of the ways in which misophonia is inadequately understood by health care researchers, but these do represent major areas in need of research attention.

In order for any of the aforementioned areas to be fully investigated, the establishment of reliable and valid assessment tools is necessary. At the present time, there are two self-report measures that have not been fully evaluated for their validity in assessing the severity of misophonia. Still necessary is a systematic clinician-administered diagnostic tool, whether as a standalone measure or integrated into existing protocols, that will definitively identify individuals with misophonia, as well as properly rule out conditions that might be primary, and for which selective sound sensitivity is a secondary problem, such as high SPS.

7.1 Couching misophonia in fundable terms

Any program of research must establish the public health relevance of the problem, and then identify ways to investigate it. Through this process, information can then be shared with health providers so that they may be equipped with reliable information and methods for alleviating symptoms, as well as ruling out comorbidities. As noted earlier, the RDoC initiative of NIMH prioritizes research that is based on valid indicators of psychopathology. Accordingly, a first priority is determining the extent that misophonia can be validly assessed. Once that is accomplished, the potential to further investigate the relevant domains outlined in the RDoC matrix could be undertaken, covering the gamut of biomedically salient features that may undergird misophonia.

More immediately, after establishing a valid assessment method for misophonia, it may be possible to evaluate the way in which it would fit in the HiTOP model. Fig. 2 depicts the core factors in HiTOP (for a more granular breakdown, see Conway et al., 2019). The placement of misophonia in this model is not known, pending development of a valid instrument for assessing the condition. However, once placed in this context, greater clarity for how to best develop treatment programs will then come into focus.

7.2 Getting the word out

Any effort to successfully understand a condition requires public awareness. Misophonia has recently garnered media attention, as well as SPS. This is helpful, and more is still necessary so individuals may identify themselves and others, and develop adaptive management strategies. The accumulation of case data is essential to start deriving clinical hypotheses regarding what is feasible to test in larger and more controlled settings. This may mean developing networks of providers with a shared background, however limited it may be, in investigating and treating misophonia. It may also involve pooling available unpublished cases into a larger systematic review to draw broader conclusions about the nature of the condition and what meaningful next steps may be taken in the research. What can be concluded now is that although we have made some progress in identifying both misophonia and SPS—states which are associated with heightened sensitivities that are mediated via brain signals—much work remains to be done not only to fully characterize them but to develop interventions and strategies that may mitigate suffering and permit these individuals to thrive in the world and with others.

References

Abramowitz, J., McKay, D., & Taylor, S. (2005). Subtypes of obsessive-compulsive disorder. *Behavior Therapy, 36*, 367–369.

Acevedo, B. P., Aron, E. N., Aron, A., Sangster, M. D., Collins, N., & Brown, L. L. (2014). The highly sensitive brain: An fMRI study of sensory processing sensitivity and response to others' emotions. *Brain & Behavior, 4*, 580–594.

Acevedo, B., Aron, E., Pospos, S., & Jessen, D. (2018). The functional highly sensitive brain: A review of the brain circuits underlying sensory processing and seemingly related disorder. *Philosophical Transactions of the Royal Society B, 373*, 20170161.

Acevedo, B. P., Jagiellowicz, J., Aron, E., Marhenke, R., & Aron, A. (2017). Sensory processing sensitivity and childhood quality's effects on neural responses to emotional stimuli. *Clinical Neuropsychiatry, 14*, 359–373.

American Psychiatric Association. (2013). *Diagnostic and statistical manual of mental disorders* (5th ed.). Arlington, VA: Author.

Barlow, D. H. (2002). *Anxiety and its disorders* (2nd ed.). New York: Guilford.

Bernstein, R. E., Angell, K. L., & Dehle, C. M. (2013). A brief course of cognitive behavioural therapy for the treatment of misophonia: A case example. *The Cognitive Behaviour Therapist, 6*, e10.

Brown, T. A., & Barlow, D. H. (2015). *Anxiety and related disorders interview schedule for DSM-5, lifetime version: client interview version*. New York: Oxford University Press.

Conway, C. C., Forbes, M. K., Forbush, K. T., Fried, E. I., Hallquist, M. N., Kotov, R., Mullins-Sweatt, S. N., … & Eaton, N. R. (2019). A hierarchical taxonomy of psychopathology can transform mental health research. *Perspectives on Psychological Science, 14*, 419–436.

Craske, M. (2003). *Origins of phobias and anxiety disorders: Why more women than men?* Amsterdam: Elsevier.

Cuthbert, B. N., & Kozak, M. J. (2013). Constructing constructs for psychopathology: The NIMH Research Domain Criteria. *Journal of Abnormal Psychology, 122*, 928–937.

Davey, G. C., & Wells, A. (2006). *Worry and its psychological disorders: Theory, assessment, and treatment*. Chichester, UK: Wiley.

Dawson, G., & Watling, R. (2000). Inerventions to facilitate auditory, visual, and motor integration in autism: A review of the evidence. *Journal of Autism and Developmental Disorders, 30*, 415–421.

Edelstein, M., Brang, D., Rouw, R., & Ramachandran, V. (2013). Misophonia: Physiological investigations and case descriptions. *Frontiers in Human Neuroscience, 7*, 296.

Fink, G. (Ed.), (2009). *Stress consequences*. Amsterdam: Academic Press.

First, M. B., Williams, J. B. W., Karg, R. S., & Spitzer, R. L. (2015). *Structured clinical interview for DSM-5: Research version*. Arlington, VA: American Psychiatric Association.

Foa, E. B., & Kozak, M. J. (1986). Emotional processing of fear: Exposure to corrective information. *Psychological Bulletin, 99*, 20–35.

Folmer, R. L., Theodoroff, S. M., Martin, W. H., & Shi, Y. (2014). Experimental, controversial, and futuristic treatments for chronic tinnitus. *Journal of the American Academy of Audiology, 25*, 106–125.

Gorka, A. X., LaBar, K. S., & Hariri, A. R. (2016). Variability in emotional responsiveness and coping style during active avoidance as a window onto psychological vulnerability to stress. *Physiology & Behavior, 158*, 90–99.

Heimberg, R. G., Turk, C. L., & Menin, D. S. (2004). *Generalized anxiety disorder: advances in research and practice*. New York: Guilford.

Insel, T., Cuthbert, B., Garvey, M., Heinssen, R., Pine, D. S., Quinn, K., … Wang, P. (2010). Research Domain Criteria (RDoC): Toward a new classification framework for research on mental disorders. *American Journal of Psychiatry, 167*, 748–751.

Jastreboff, M. M., & Jastreboff, P. J. (2002). Decreased sound tolerance and Tinnitus Retraining Therapy (TRT). *Australian and New Zealand Journal of Audiology, 24*, 74–81.

Kazdin, A. E. (2007). Mediators and mechanisms of change in psychotherapy research. *Annual Review of Clinical Psychology*, *3*, 1–27.
Krueger, R. F., Caspi, A., Moffitt, T. E., & Silva, P. A. (1998). The structure and stability of common mental disorders (DSM-III-R): A longitudinal-epidemiological study. *Journal of Abnormal Psychology*, *107*, 216–227.
Lazarus, R. S., & Folkman, S. (1984). *Stress, appraisal, and coping*. New York: Springer.
Lilienfeld, S. O. (2014). The Research Domain Criteria (RDoC): An analysis of methodological and conceptual challenges. *Behaviour Research and Therapy*, *62*, 129–139.
McGuire, J. F., Wu, M. S., & Storch, E. A. (2015). Cognitive-behavioral therapy for 2 youths with misophonia. *Journal of Clinical Psychiatry*, *76*, 573.
McKay, D. (2017). Presidential address: Embracing the repulsive: the case for disgust as a functionally central emotional state in the theory, practice, and dissemination of cognitive-behavior therapy. *Behavior Therapy*, *48*, 731–738.
McKay, D., Kim, S. K., Mancusi, L., Storch, E. A., & Spankovich, C. (2018). Profile analysis of psychological symptoms associated with misophonia: A community sample. *Behavior Therapy*, *49*, 286–294.
McKay, D., & Storch, E. A. (2018). *Misophonia severity index*. Unpublished scale Department of Psychology, Compulsive, Obsessive, and Anxiety Program (COAP), Fordham University.
Miloyan, B., & Doorn, G. (2019). Longitudinal association between social anxiety disorder and incident alcohol use disorder: Results from two national samples of US adults. *Social Psychiatry & Psychiatric Epidemiology*, *54*, 469–475.
Mowrer, O. H. (1960). *Learning theory and behavior*. New York: Wiley.
Olatunji, B. O., & Broman-Fulks, J. J. (2007). A taxometric study of the latent structure of disgust sensitivity: Converging evidence for dimensionality. *Psychological Assessment*, *19*, 437–448.
Olatunji, B. O., & Wolitzky-Taylor, K. B. (2009). Anxiety sensitivity and the anxiety disorders: A meta-analytic review and synthesis. *Psychological Bulletin*, *135*, 974–999.
Palumbo, D. B., Alsalman, O., de Ridder, D., Song, J. J., & Vanneste, S. (2018). Misophonia and potential underlying mechanisms: A perspective. *Frontiers in Psychology*, *9*, 953.
Pilecki, B. C., Clegg, J. W., & McKay, D. (2011). The influence of corporate and political interests on models of illness in the evolution of the DSM. *European Psychiatry*, *26*, 194–200.
Reid, A. M., Guzick, A. G., Germand, A., & Olsen, B. (2016). Intensive cognitive-behavioral therapy for comorbid misophonic and obsessive-compulsive symptoms: A systematic case study. *Journal of Obsessive-Compulsive and Related Disorders*, *10*(1–9).
Reiss, S., Peterson, R. A., Gursky, D. M., & McNally, R. J. (1986). Anxiety sensitivity, anxiety frequency and the predictions of fearfulness. *Behaviour Research and Therapy*, *24*, 1–8.
Rozin, P., & Fallon, A. E. (1987). A perspective on disgust. *Psychological Review*, *94*, 23–41.
Rozin, P., Haidt, J., & Fincher, K. (2009). From oral to moral. *Science*, *323*, 1179–1180.
Schröder, A., Vulink, N., & Denys, D. (2013). Misophonia: Diagnostic criteria for a new psychiatric disorder. *PLOS One*, *8*, e54706.
Schröder, A. E., Vulink, N. C., van Loon, A. J., & Denys, D. A. (2017). Cognitive behavioral therapy is effective in misophonia: An open trial. *Journal of Affective Disorders*, *217*, 289–294.
Sergeant, J. (2000). The cognitive-energetic model: An empirical approach to Attention-Deficit Hyperactivity Disorder. *Neuroscience and Biobehavioral Reviews*, *24*, 7–12.
Skinner, B. F. (1938). *The behavior of organisms*. New York: Appleton-Century-Kroft.
Stein, D. J., Phillips, K. A., Bolton, D., & Fulford, K. W. M. (2010). What is a mental/psychiatric disorder? From DSM-IV to DSM-V. *Psychological Medicine*, *40*, 1759–1765.
Stiegler, L. N., & Davis, R. (2010). Understanding sound sensitivity in individuals with autism spectrum disorders. *Focus on Autism and Other Developmental Disabilities*, *25*, 67–75.
Taylor, S. (1999). *Anxiety sensitivity*. Mahwah, NJ: Erlbaum.

Taylor, S. (2017). Misophonia: A new mental disorder? *Medical Hypotheses*, *103*, 109–117.
Taylor, S., Asmundson, G. J. G., Abramowitz, J. S., & McKay, D. (2009). Classification of anxiety disorders for DSM-V and ICD-11: Issues, proposals, and controversies. In D. McKay, J. S. Abramowitz, S. Taylor, & G. J. G. Asmundson (Eds.), *Current perspectives on anxiety disorders: Implications for DSM-V and beyond* (pp. 481–511). New York: Springer.
Waller, N. G., & Meehl, P. E. (1998). *Multivariate taxometric procedures: Distinguishing types from continua*. Newbury Park, CA: Sage.
World Health Organization. (2015). *International classification of diseases* (10th ed.). Geneva, Switzerland: Author.
Wu, M. S., Lewin, A. B., Murphy, T. K., & Storch, E. A. (2014). Misophonia: Incidence, phenomenology, and clinical correlates in an undergraduate student sample. *Journal of Clinical Psychology*, *70*, 994–1007.
Zhou, X., Wu, M. S., & Storch, E. A. (2017). Misophonia symptoms among Chinese university students: Incidence, associated impairment, and clinical correlates. *Journal of Obsessive-Compulsive and Related Disorders*, *14*, 7–12.

CHAPTER 8

The future of sensory processing sensitivity on planet Earth and beyond

Bianca P. Acevedo
Neuroscience Research Institute, University of California, Santa Barbara, CA, United States

Contents

1 The future of sensory processing sensitivity beyond planet Earth 188
1.1 The challenges of space travel for SPS 189
1.2 The benefits of SPS for space travel 194
References 195

> *Empathic awareness allows us to experience others' suffering or joy as our own, but also, it loops back to help reinforce our sense of selfhood as we come to see strands of others in us, and a bit of ourselves in others.*
>
> ***Jeremy Rifkin***

In the winter of 2019 I started getting an increasing number of requests from groups around the world launching global initiatives to raise awareness about sensory processing sensitivity (SPS), or as termed in my native Spanish tongue, "La Alta Sensibilidad." At the time, I had started a new position as Director of Research for an organization running treatment centers for individuals with opioid use disorders in New York City. Needless to say, after moving cross-country, starting a new job, finishing a book, and being highly sensitive myself, I was feeling overwhelmed. Thus, I had to filter many of the requests to speak about high sensitivity, and even had to say "no" to some, in order to maintain by equilibrium and complete my various commitments. Perhaps this has been a strategy that has served me, even to others' discontent, that I do slow down and take time for myself so I don't burn out. This is especially important for highly sensitive persons (HSPs).

Out of the blue, in the early spring 2020, I received a curious request from a team of researchers writing about human missions to Mars, the Moon, and off-world settlement. They contacted me to gauge my interest in writing a book chapter with a psychological perspective about space travel and human exploration of Luna (Earth's Moon) with a "*human factor*"

The Highly Sensitive Brain
https://doi.org/10.1016/B978-0-12-818251-2.00008-4

spin. Although I was already stretched thin for time, and I realized that this project would take me into new territory (no pun intended); requiring me to read, research, and interview people in fields outside of my expertise, my interest was so piqued that I could *not* say "no."

With regard to space travel, the human factor explores questions and solutions about how humans might adapt to the challenges of space flight, exploration, and settlement. For example, confinement, isolation, monotony, need for exercise and meaningful activity, movement in microgravity, limited resources, exposure to radiation, and artificially created dry air are just some of the conditions that humans may experience in space. These conditions certainly will challenge people's psychological, physical, emotional, and cognitive states. Thus, the human factor asks questions such as how will humans find ways to stay motivated, play, and work during long trips of space travel where many days may be spent onboard a spacecraft with fellow crew members that are neither kin, nor a spouse, nor a close other. How will humans socialize and deal with common issues that arise, such as conflicts, need for space, mating, and claiming of territory? Also, it is likely that in the course of space travel and beyond, other normal emotions and thought patterns may arise such as missing loved ones, pets, and favorite Earthling things left behind. Thus, how to deal with both mild and intense emotions—ranging from calm and fatigue, to fear, anxiety, grief, and stress—will pose another challenge, especially given the limited social support and outlets for releasing tension on space missions.

However, solutions such as round-the-clock virtual experts have been proposed. This of course has the constraint of being able to connect with these experts, which may not always work if there are signal failures. Also, solutions to outer-world problems proposed on Earth may not apply in outer space as humans' physiology may become so affected by extreme conditions in outer planets that basic processes which occur naturally and usually go unnoticed—such as breathing, sleeping, loving, empathizing, and the desire to mate ——may become dysregulated or impaired. Thus, beyond dealing with practical issues, changes to our physiology may impact the basic psychological, emotional, attachment-related fabric of our species. These changes will certainly have a stronger impact on the highly sensitive brain and system.

1 The future of sensory processing sensitivity beyond planet Earth

Although space exploration may be stretching far into the future, it is not *im*possible and it's in our near future as several private companies, including

Elon Musk's SpaceX, have proposed to launch space tourism missions to International Space Stations by 2021 (e.g., CNN: Wattles, 2020). Scientists, engineers, and administrators are considering piloted missions and colonization of Mars and other extraterrestrial bodies such as Luna, Earth's moon (NASA, 2014, 2016; Szocik, Wójtowicz, Boone Rappsaport, & Corbally, 2020). Also, there have been ongoing studies, plans, and thinking about space travel for over a few decades now, including simulations of extreme environments here on Earth. For example, projects such as Biosphere 2 (a laboratory built in Arizona to examine global ecological processes, which was inhabited by eight humans for 2 years, approximately the same amount of time for a human space flight to Mars) and expeditions to Antarctica provide a window into extreme environments. These human expeditions have shed light on some of the experiences and challenges that colonists may face in space such as isolation, and lack of adequate food, air, and light (Nelson, 2018; Smith, Kinnafick, & Saunders, 2017).

Why might pondering about SPS in outer space make sense? Given that HSPs are more affected by their environments, both for better and for worse, they will manifest symptoms more readily compared to those less sensitive. Also, HSPs are more likely to show rapid adaptations to their environments. This of course although seemingly negative at times may serve their survival in the long run. Nevertheless, for highly sensitive individuals, the challenges of inhabiting extreme habitats with limited resources and members that may be less sensitive or experiencing issues themselves may be far greater. Thus, I will start my discussion on the future of SPS in outer space and work my back to planet Earth.

1.1 The challenges of space travel for SPS

As mentioned previously, there are many trials and tribulations that humans may face in space travel, and HSPs with their highly sensitive brains and systems, are likely to experience such events more strongly. To start, "time in solitude" may be nearly impossible on a spacecraft or outer planet settlements. This poses a special challenge for HSPs, which require "alone-time" to recover from overstimulation stirred by sensory stimuli including others' emotions.

Indeed, greater empathy is a cardinal feature of SPS, which has been established both with self-report questionnaires (Aron & Aron, 1997) and brain imaging techniques (Acevedo, Aron, & Aron, 2014). As described in previous chapters of this volume, our team showed that SPS was associated with stronger neural activity in areas such as the inferior frontal gyrus (IFG), known for having "mirror neurons" and playing a key role in empathic processing (Acevedo et al., 2014). Empathy is a necessary component for perceiving and responding to another person's needs.

Given the extreme conditions of space travel and off-world settlement, it is likely that every crew member will, at some point, experience both mild and intense emotional, psychological, physical, and cognitive difficulties. HSPs will certainly be more affected by others' issues, big and small. At best, the HSP may feel compassion for their fellow crew member's suffering. If overwhelmed, they may become distressed, reclusive, shutdown, and isolate themselves (which certainly will be challenging with limited space) as a means to not absorb negativity from others. At worst, the highly sensitive individual may themselves come to experience complications, symptoms, and may even fall ill.

However, as described in previous chapters, having a positive childhood is an important factor in creating resiliency and long-term adaptive outcomes for HSPs (Liss, Timmel, Baxley, & Killingsworth, 2005). Thus, it will be important to screen for having had a positive childhood for all individuals initially, but particularly among the highly sensitive. Indeed, research shows that HSPs with positive childhoods are resilient to the effects of negative stimuli, including others' distressing emotions (Acevedo et al., 2017). As noted in previous chapters, in one brain imaging study my team members and I found that high SPS individuals showed decreased reward activation in response to negative images, such as pictures of an accident or a dead animal. However, when we accounted for participants' positive childhoods, the typical diminished reward response to negative images was no longer apparent. Instead, HSPs with positive childhoods showed more activation of brain areas related to self-regulation, feelings of calmness, and physiological homeostasis when viewing distressing pictures (Acevedo, Aron, Pospos, & Jessen, 2018). The ability to stay calm under stress and resiliency are qualities that may be especially sought after for long space missions with new and many challenges. Thus, HSPs with positive childhoods may end up being prime candidates for these space missions, as they may have enhanced resiliency and the ability to stay calm and physiologically self-composed in the face of stress.

Empathy is another human factor that is important to consider, and possibly screen for, when recruiting (low, moderate, or highly sensitive) individuals for space missions. As noted previously, conditions in extreme environments with limited social support require team work. Thus, it will be vital for the success of space missions for the crew to be sensitive to one another's needs and respond appropriately. Indeed, there are both adaptive and maladaptive types of empathy, which researchers have delineated and measure with self-report scales. One popular measure, the Interpersonal Reactivity Index (IRI; Davis, 1980), defines empathy as the "reactions of one individual to the observed experiences of another." This 28-item scale measures facets of empathy, including perspective taking (PT), empathic concern (EC), and personal distress (PD). PD, which is the least adaptive types of empathy, assesses "self-oriented" feelings of personal anxiety and unease in tense interpersonal situations and in response to another's suffering. Although it may seem kind to respond to another's suffering by experiencing their pain, it is not functional because it inhibits appropriate responsiveness. Also, individuals with high PD may not wish to experience discomfort and anxiety, so they may develop coping strategies such as avoiding individuals in that are suffering or in distress. Avoidance toward strangers in distress may not have significant repercussions for the individual (although it contradicts altruism); however, it can create serious issues for attachment relationships, especially toward children and close others, as responsive support is a basic tenet of healthy attachment relationships (Bowlby, 1969; Pietromonico & Collins, 2017). Thus, having a robust and adaptive empathy style will be beneficial on space missions and at large.

In space missions crew members may become attached and will need to rely on one another for many things, including healthy and useful support. Given HSPs' capacity for empathy, responsiveness, and calmness (if they are not burnt out themselves), high SPS individuals may provide an invaluable resource to the group in times of stress. This is important to keep in mind and harness going forward.

Everyday social processes and more complicated types of attachment bonding behaviors, which are so deeply embedded in our evolutionary wiring, physiology, and psyche, will play a significant role in space travel and off-world settlements. What types of social challenges might our species face during space missions? Clues about the social challenges that our species might face when spending time in space travel and colonizing new worlds are provided from our ancestral history. For example, issues such as violence, aggression, war, fights over territory and mates, in-group/out-group conflicts, dominance, and submission did and do exist. However, other milder perspectives have been offered to explain how our ancestors spent their time and how they socialized.

Relatively recent thinking regarding the nature of our ancestors' temperament provided a paradigm shift for scientists and thinkers (Rifkin, 2009).

Contrary to the savage, competitive warrior depiction of humans, some archeologists and historians have proposed that widespread violence has not been the norm in human history if we consider the entire span that anatomically modern human beings have existed on Earth (Gimbutas, 1991). (Although it may be hard to imagine, especially if we have just opened our news browser or glanced at the front page of the newspaper given the bias toward negativity and sensationalism.) Indeed, for 93% of our species' existence, humans lived as foragers and hunters in small tribal groups consisting of about 30–150 people. These archaic people were nomadic and communal. While aggression and violence existed among some, it was generally limited in scale and confined to protecting territory against intruders and conflicts over mate selection. It is thought that humans spent far more time grooming, foraging, playing, and socializing rather than fighting and killing each other.

For example, in the early European garden/agricultural societies of the Neolithic Age archeologists have found very little evidence of weapons, but instead found pottery and art (Gimbutas, 1991). Thus, some thinkers have proposed that early European agricultural societies may have been largely peaceful, egalitarian, matrilineal, and artistic (Gimbutas, 1991; Keller, 1996). This reframing of our ancestors as peaceful and cooperative rather than aggressive warriors has been a paradigm shift. Emerging evidence on the biological basis of altruism, sociality, and attachment, including toward nonkin, have further contributed to our understanding of the roots of our complex social behaviors including cooperation, mating strategies (both short and long term), monogamy, and even altruism toward strangers. As a result of new evidence, many scientists, philosophers, theologians, and historians have revised their thinking to consider integrated views of humanity, including interdependence, cooperation over competition, and the relationship of humans to the larger cosmic whole (Keller, 1996, p. 88).

The new science of positive psychology and empirical study of our SPS systems have resulted in scientifically valid ways to assess these factors in humans. For example, "Inclusion of Other in the Self" theory, with its corresponding one-item Venn diagram measure, assesses the extent to which individuals perceive a sense of cognitive overlap or interdependence with another individual or entity (Aron, Aron, & Smollan, 1992). Thus, this measure also may be useful in screening for space missions as it will be important for each member to feel that they are a part of and integrated with the group, as well as the cosmic whole. This will facilitate space crews in adapting to the new group and off-world environment.

Mind-body techniques, such as yoga and meditation, may be useful during space travel as these practices facilitate bringing the human brain and system to equilibrium without the use of external props or substances. Indeed, there is a growing body of work showing the many neural benefits of mind-body practices (Acevedo, Pospos, & Lavretsky, 2016). For example, studies have shown that engagement in mind-body techniques significantly reduce stress, anxiety, and depression (Davis, Hayes, & Hilsenroth, 2011; Irving, Dobkin, & Park, 2009; Kabat-Zinn, 2003). Also, practicing yoga and meditation has been shown to result in better mood, enhanced cognitive functioning, and increased empathy and resiliency (Birnie, Speca, & Carlson, 2010; Eyre, Siddharth, Acevedo, & Lavretsky, 2017).

For example, my colleagues at the University of California, Santa Barbara, and I found that following a 6-week mind-body intervention, participants showed significant reductions in Empathic PD measured with the IRI (Acevedo & Santander, 2018). Moreover, decreases in PD were associated with increases in general quality of life, and enhanced relationship satisfaction, sex satisfaction, and feelings of love for the partner (Acevedo, Santander, & Kosik, 2019). Thus, a simple 3-minute meditation practiced everyday may help individuals to have better resiliency (with lower empathic PD) and improved relationship factors. This type of technique may be especially beneficial for highly sensitive individuals and spaceflight members, helping them to feel more connected and buffering them from the social challenges that they may encounter during space travel and off-world settlement.

The study also showed that decreases in PD were associated with widespread patterns of stronger resting-state neural connectivity across nearly the whole brain, with the strongest effects shown in the ventral attention network and subsystems of the default mode network (involved in emotional and self-referential processing). We also observed associations between brain connectivity with PT and EC, which were again largely driven by connectivity in executive/emotional control networks and components of attention and default mode systems. Thus, a simple meditation technique, in addition to reducing PD, also strengthens brain connectivity.

These findings suggest that mind-body practices, specifically meditation, may be helpful during space travel for strengthening whole-brain connectivity. This is a fruitful area for further investigation, which may possibly help humans in space to strengthen their brain connectivity and therefore affect the entire human physiological system. Also, other types of meditations may be explored, which might strengthen other systems of the brain as the one described here mainly targeted systems implicated in emotion, self-regulation, and self-reflective capabilities.

1.2 The benefits of SPS for space travel

Thus far we have discussed some of the challenges that humans, specifically highly sensitive ones, may encounter in space travel and missions. However, there may be benefits to being highly sensitive for adapting to novel and extreme conditions in space. For example, although high SPS individuals tend to feel things more strongly, they may have an easier time adapting to new conditions and extreme environments, as they have learned to do so on planet Earth. Most HSPs may have developed coping strategies for facing what they feel to be "extreme conditions" already. Some strategies are more adaptive than others. Second, the solitude and monotony of space flight and settlement may not bother HSPs as much as less sensitive individuals. Third, somewhat muted sensory experiences, such as lower sound frequencies and limited light, may be more welcome by HSPs, who commonly report preferring minimalistic, low sensory contexts to "recover" from everyday overstimulation. Also, inhabiting a new planet with fewer people, things, and sources of overstimulation may not be the worst thing for HSPs, as many tend to feel overwhelmed on planet Earth. A minimalistic planet with a simpler diet and less chaotic environment may not be an ideal environment for a low sensation-seeking HSP. It's even possible that this new environment may have a profound effect on the highly sensitive brain and system creating feelings of connection, awe, calm, and wonder.

As noted previously, SPS is characterized by greater awareness of subtleties in the environment. Thus, HSPs may use their attunement to aesthetics and subtleties for creative endeavors, including the arrangement of their environments to enhance comfort and ease for themselves and others. Being highly conscientious, high SPS persons are also the type that take up causes, feel deeply for others, care for the well-being of the environment; and thus, they will be invaluable in establishing rules and systems in outer-world settlements. Also, as mentioned in this volume, when highly sensitive individuals have had a positive childhood (and they themselves are not feeling overwhelmed) they may display enormous empathy and responsiveness to others' needs. These qualities will certainly be invaluable in any stressful situation, especially space travel and off-world settlement.

In sum, HSPs have a lot to offer in space missions and also outer-planet settlements. But why wait? We have much to do on planet Earth. Not necessarily to modify the world so that it suits our needs, preferences, and longings; or varying levels of sensitivity but to enhance sustainability of the planet and quality of life of life for all. However, sensitivity can be used as a guide to ensure that conditions are hospitable, comfortable, and beneficial for all. It is important to be aware that different ways of being exist and not to expect that all individuals will respond to things the same way. Also, to harness the talents of the highly sensitive is another noble goal. Finally, to fight for all children's well-being is critical for creating a future society that we would all be happy, proud, and comfortable to inhabit.

Thus, looking toward a future society, whether here on Earth or beyond, it's imperative to give a concerted effort to creating conditions where cooperation over competition, peace over war, respect for nature, egalitarian bonds, human creativity, and ecological sustainability prevail. In sum, many of the qualities that highly sensitive individuals engender have been suggested to be central to our ancestral heritage and hopefully also for the future. Because it is believed that, once upon a time, humans lived by these values, these have been suggested as a guide for the future societies on planet Earth, but perhaps also elsewhere.

References

Acevedo, B., Aron, E., Pospos, S., & Jessen, D. (2018). The functional highly sensitive brain: A review of the brain circuits underlying sensory processing sensitivity and seemingly related disorders. *Philosophical Transactions of the Royal Society B*, https://doi.org/10.1098/rstb.2017.0161.

Acevedo, B., Jagiellowicz, J., Aron, E., Aron, A., Marhenke, R., & Aron, A. (2017). Sensory processing sensitivity and childhood quality's effects on neural responses to emotional stimuli. *Clinical Neuropsychiatry*, *14*, 359–373.

Acevedo, B., & Santander, T. (2018). Yoga, empathy, and the human brain: A pilot study. In *International Association for Yoga therapy, symposium for yoga research, Kripalu Institute, Berkshirees, MA, October 15–17, 2018*.

Acevedo, B., Santander, T., & Kosik, K. (2019). Yoga, empathy, and brain connectivity in caregivers. In *Alzheimer's Association International conference, Los Angeles, CA, USA, July 14–17, 2019*.

Acevedo, B. P., Aron, E. N., & Aron, A. (2014). The highly sensitive brain: An fMRI study of sensory processing sensitivity and response to others' emotions. *Brain and Behavior*, *4*(4), 580–594.

Acevedo, B. P., Pospos, S., & Lavretsky, H. (2016). The neural mechanisms of meditative practices: Novel approaches for healthy aging. *Current Behavioral Neuroscience Reports*, *3*(4), 328–339.

Aron, A., Aron, E. N., & Smollan, D. (1992). Inclusion of other in the self scale and the structure of interpersonal closeness. *Journal of Personality and Social Psychology*, *63*, 596–612. https://doi.org/10.1037/0022-3514.63.4.596.

Aron, E. N., & Aron, A. (1997). Sensory-processing sensitivity and its relation to introversion and emotionality. *Journal of Personality and Social Psychology*, *73*(2), 345–368.

Birnie, K., Speca, M., & Carlson, L. (2010). Exploring self-compassion and empathy in the context of mindfulness-based stress reduction (MBSR). *Stress and Health*, *26*(5), 359–371.

Bowlby, J. (1969). *Attachment and Loss, Vol. 1, Attachment*. London: Hogarth.

Davis, M. H. (1980). A multidimensional approach to individual differences in empathy. *JSAS Catalog of Selected Documents in Psychology*, *10*, 85.
Davis, D., Hayes, J., & Hilsenroth, M. J. (2011). What are the benefits of mindfulness? A practice review of psychotherapy-related research. *Psychotherapy*, *48*(2), 198–208.
Eyre, H. A., Siddharth, P., Acevedo, B., & Lavretsky, H. (2017). A randomized control trial of Kundalini yoga in mild cognitive impairment. *International Psychogeriatrics*, https://doi.org/10.1017/S1041610216002155.
Gimbutas, M. (1991). *The civilization of the Goddess: The world of old Europe*. San Francisco: Harper.
Irving, J. A., Dobkin, P. L., & Park, J. (2009). Cultivating mindfulness in health care professionals: A review of empirical studies of mindfulness-based stress reduction (MBSR). *Complementary Therapies in Clinical Practice*, *15*(2), 61–66.
Kabat-Zinn, J. (2003). Mindfulness-based interventions in context: Past, present, and future. *Clinical Psychology: Science and Practice*, *10*(2), 144–156.
Keller, M. (1996). Gimbutas's theory of early European origins and the contemporary transformation of Western civilization. *Journal of Feminist Studies in Religion*, *12*(2), 73–90.
Liss, M., Timmel, L., Baxley, K., & Killingsworth, P. (2005). Sensory processing sensitivity and its relation to parental bonding, anxiety, and depression. *Personality and Individual Differences*, *39*, 1429–1439. https://doi.org/10.1016/j.paid.2005.05.007.
NASA, National Aeronautics and Space Administration. (2014). *NASA's journey to mars*. Retrieved from https://www.nasa.gov/content/nasas-journey-to-mars.
NASA, National Aeronautics and Space Administration. (2016). *Mars 2020 mission overview*. Retrieved from https://mars.nasa.gov/mars2020/mission/overview/.
Nelson, M. (2018). *Pushing our limits: Insights from Biosphere 2*. Tucson: The University of Arizona Press.
Pietromonico, P., & Collins, N. (2017). Interpersonal mechanisms linking close relationships to health. *American Psychologist*, *72*, 531–542.
Rifkin, J. (2009). *The Empathic Civilization: The race to global consciousness in a world in crisis*. New York: Penguin.
Smith, N., Kinnafick, F., & Saunders, B. (2017). Coping strategies used during an extreme Antarctic expedition. *Journal of Human Performance in Extreme Environments*, *13*, 1–10. https://doi.org/10.7771/2327-2937.1078.
Szocik, K., Wójtowicz, T., Boone Rappaport, M., & Corbally, C. (2020). Ethical issues of human enhancements for space missions to Mars and beyond. *Futures*, *115*, 1–14.
Wattles, J. (2020). *SpaceX signs deal to send tourists and others to International Space Station*. CNN. Retrieved from https://www.cnn.com/2020/03/05/tech/spacex-tourism-passengers-iss/index.html.

APPENDIX

Misophonia severity index

Dean McKay[a], Eric A. Storch[b]
[a]Department of Psychology, Fordham University, Bronx, NY, United States
[b]Baylor College of Medicine, Compulsive, Obsessive, and Anxiety Program (COAP) Lab, Fordham University, Bronx, NY, United States

1 Symptom checklist

Many misophonia sufferers report distress from a wide range of "trigger sounds," which set in motion a wide range of emotional reactions. Please note below which of the sounds below have caused distress in the past but do not any longer, or currently cause distress. For the symptoms that *currently* cause distress, please rate the severity using the following scale:

1 = minimal
2 = moderate
3 = severe
4 = extreme/disabling

Please check column		Trigger sound	Please check column for current distress			
Current	**Past**		**1 Minimal**	**2 Moderate**	**3 Severe**	**4 Extreme/Disabling**
Mouth/throat sounds						
		Slurping				
		Chewing				
		Crunching sounds				
		Brushing teeth				
		Throat clearing				
		Coughing (with mucous)				
		Lip smacking				
		Gulping after drinking fluid				
		Audible "ahh" sound following swallowing				
		Spitting				
		Sucking sounds (i.e., on candy)				
		Gum chewing/cracking				
		Kissing sounds				
Breathing/respiratory sounds						
		Loud breathing				
		Shortness of breath sounds				
		Grunting				
		Yawning				
		Sniffling				
		Raspy voices				
		Nose whistling				
		Hiccups				
Vocal sounds						
		Humming				
		Whistling				

		Singing Soft whispering Muffled talking Hard consonants Nasal voices				
Body movement sounds						
		Foot or hand tapping Finger snapping Cracking knuckles Nail biting/clipping Foot shuffling				
Other sounds						
		Clicking sounds from texting, typing, mouse clicks Cell phone sounds (ringtones, other noises) Wrappers (plastic bags, hard plastic wrap, cellophane) Lawnmowers, leafblowers, refrigerator compressor, air conditioner Dogs barking Babies crying Birds chirping Animals scratching or licking or objects selves Basketball thumps Silverware scraping (i.e., on a plate or bowl)				
Other sounds not listed above						

Clinician administered items:

Considering the current symptoms noted in the checklist, rate the following items:

1. How many times per day do you find that sounds cause you emotional distress?
   (1) infrequently, possibly one or two times a day sounds will bother me
   (2) regularly, sounds bother me several times a day
   (3) frequently, sounds are often intruding on my daily life
   (4) near constant, sounds bother me almost all the time
2. When sounds cause you emotional distress, how much interference do these sounds cause?
   (1) mild, I can generally still complete activities
   (2) moderate, it interferes with activities but I can still do some easy tasks
   (3) severe, I have trouble doing even basic tasks
   (4) extreme, I cannot do anything and have to wait for the distress to pass
3. How much do you avoid situations and/or people due to the sounds?
   (1) mild, sounds do not interfere with my daily activities or social life
   (2) moderate, sounds interfere with some activities and/or there are some people I make a point of avoiding
   (3) severe, sounds interfere with most of my activities, and/or there are numerous people I avoid because of noise
   (4) extreme, sounds interfere with virtually all activities and/or has completely interfered with my social life
4. What specific emotional reactions do you experience when confronted with the sounds? (choose all that apply)
   a. Anxiety
   b. Anger
   c. Disgust
   d. Dysphoria

Index

Note: Page numbers followed by *f* indicate figures and *t* indicate tables.

A

Active self-regulation, 80–81
Acute stress disorder (ASD), 177
Adolescents, health and social outcomes
  anxiety, 76–80
  depression, 76–80
  peer relationships, 86–88
  social adjustment, moderators of, 80–86
Adrenocorticotropic hormone (ACTH), 125
Adults, health and social outcomes
  life satisfaction and happiness, 92–93
  mental health, 89–92
  physical health, 89
  social outcomes, 95–97
  work satisfaction, highly sensitive adults, 93–94
Aesthetic sensitivity (AES), 23–25, 33–34, 42, 91, 93–94
Agreeableness, 113–114
"Altruistic" traits, 8
Amsterdam Misophonia Scale (A-MISO-S), 170–171
Anger, 173–174
Anxiety and Related Disorders Interview Schedule for DSM-5 (ARDIS-5), 172
Anxiety disorders, misophonia
  anxiety and, 174
  conceptualization of, 175
  generalized anxiety disorder (GAD), 176
  obsessive-compulsive and related disorders (OCRDs), 177
  social anxiety disorder (SAD), 176–177
Attention-deficit hyperactivity disorder (ADHD), 61–63
Attention-shifting, 80–81
Autism spectrum disorder (ASD), 12–13, 61–63
Autoradiography, 124–125
Avoidant personality disorder (APD), 91–92

B

Behavioral activation system (BAS), 20
Behavioral inhibition system (BIS), 20
Big 5 Personality scale, 19–20, 59, 92
Biological sensitivity to context (BSC) theory, 53–54, 62–63
Blood oxygenation level signals (BOLD signal), 118
Body dysmorphic disorder (BDD), 177
Borderline personality disorder (BPD), 91–92
Brain connectivity, 193–194
Brain-derived neurotrophic factor (BDNF), 116–117
Brain-wide analyses, 123

C

Central nervous system (CNS), 57
CHAOS Scale, 93
Child behavioral inhibition, 83–84
Chronic adjustment disorder (CAD), 177
Cimbi database, 113
Cognitive behavioral therapy (CBT), 88, 118
Cognitive Bias (CogBIAS) hypothesis, 59–60, 69
Cognitive reactivity, 90
Communication, social outcomes, 96–97
Conditional adaptation, 54–55
Conditioned response (CR), 166
Conditioned stimuli (CSs), 166
Context-dependent tasks, 119
Context-independent tasks, 119
Contradictory traits, 139–140
Cooperation, 8, 192
Corticotropin-releasing hormone (CRH), 125
Creativity, 92
Criticism, sensitivity, 153–156

D

Default mode network, 193
Depression, health and social outcomes, 76–80

Developmental plasticity, 58
Developmental psychology theory, 20
Diathesis-stress model, 18–19, 52–53, 63–65
Differential susceptibility (DS) theory, 18–21, 53–55, 65–67, 140–142
Difficult temperament, 18–19, 52–53
Disgust, 173–174
Dopamine, 114–116

E

Ease of excitation (EOE), 23–25, 33–34, 42, 89, 91, 93–94
Effortful control, 80–81
Elon Musk's SpaceX, 188–189
Emotional control networks, 193
Emotional regulation, 87, 112, 152–156, 179–180
Emotion processing, 111–112, 120–122
Emotion socialization, 87
Empathic concern (EC), 190–191
Empathy, 8, 120–121, 189–191
Endophenotypic attributes, 110
Environmental sensitivity (ES), xiii, 19, 75–76, 93, 95–96
  biological sensitivity to context (BSC) theory, 53–54
  diathesis-stress model, 52–53, 63–65
  differential susceptibility (DS) theory, 54–55, 65–67
  implications, 68–70
  metaframework, 18–22, 56–58
  neurobiological and genetic basis of SPS, 67
  vantage sensitivity (VS) theory, 55, 65–67
Epigenetics, 111–112
Ethnicity, HSPs clients, 147–148
Excoriation disorder (ED), 177
"Experience sampling", 67–68
Exposure therapy, 181
Externalizing behaviors, 88

F

Facial electromyography activity, 43
Fear-conditioning test, 123–124
Freezing, 122–123
Functional magnetic resonance imaging (fMRI), 118–120, 122

G

Galvanic skin response (GSR), 179–180
Gap crossing task, 124–125
Gender, HSPs clients, 147–148
Generalized anxiety disorder (GAD), 176
Genome-wide association study (GWAS), 117–118, 126–127
Glucocorticoids, 125
Glucose metabolism, 126
Group categorization approach, 41

H

Harsh parenting, 18–19
Health and social outcomes, HSP
  in adults
    life satisfaction and happiness, 92–93
    mental health, 89–92
    physical health, 89
    social outcomes, 95–97
    work satisfaction, highly sensitive adults, 93–94
  in children and adolescents
    depression and anxiety, 76–80
    peer relationships and adolescents, 86–88
    social adjustment, moderators of, 80–86
Heightened stress system reactivity, 52–53
Hierarchical Taxonomy of Psychopathology (HiTOP), 169–170, 170*f*, 182
Highly sensitive child rating system (HSC-RS), 28–30, 29*t*, 79–80
Highly sensitive persons (HSPs), xiii–xiv, 5–7, 17–18, 51–52, 109–110, 114–115, 121, 135–138
  advantages of, 9–10
  clients, working with
    culture, 147–148
    ethnicity, 147–148
    five goals, 146–147
    gender, 147–148
    nonsensitive therapist, 143–145
    therapist and clients, 145–146
  disorder, 10–14
  environmental sensitivity (ES), 75–76
  health and social outcomes
    in adults, 88–97
    in children and adolescents, 76–88

minority, 8–9
negative coping strategies, 12
Planet Earth, 190
aesthetics and subtleties, 194
coping strategies, 194
with positive childhoods, 190
in space missions and outer-planet settlements, 194–195
Hoarding disorder (HD), 177
Human enhancements, 14
Hyper-responsiveness, 12–13
Hypothalamus-pituitary (HPA)-axis, 125

I

Immediate reactivity, 58
Inferior frontal gyrus (IFG), 189–190
Inhibitory control, 80–81
Interindividual variation, 58
International Affective Picture System (IAPS), 121
Interpersonal Reactivity Index (IRI), 190–191
Intrusive mothering behavior, 84–85
Invasive brain analyses, 123
12-item Highly Sensitive Child (HSC) self-report scale, 23–25
27-item Highly Sensitive Person scale, 22, 30–32, 32–33*t*, 35

J

Job Demands Resources Model, 93–94

L

Laboratory temperament assessment battery procedure (Lab-TAB), 28–29
Law of contagion, 173
Law of similarity, 173
Life satisfaction, happiness and, 92–93
Lifestyle, 147
Low sensory threshold (LST), 23–25, 33–34, 42, 89, 91, 93–94

M

Mental consciousness, 4–5
Mental health
anxiety and depression, 90–91
cognitive functioning, 89–90
creativity, 92
neural plasticity, 89–90
neuroticism, 90
personality disorders, 91–92
Mind-body techniques, 193
Misophonia Questionnaire (MQ), 171–172
Misophonia Severity Index (MSI), 172, 197–200, 198–199*t*
Misophonia, SPS, 165
anxiety disorders
anxiety and, 174
conceptualization of, 175
generalized anxiety disorder (GAD), 176
obsessive-compulsive and related disorders (OCRDs), 177
social anxiety disorder (SAD), 176–177
assessment of, 170–172
attention deficit disorder (with/without hyperactivity), 178
auditory-only stimuli, 179–180
auditory processing, 167
autism spectrum disorders and, 178
aversive nonmisophonic sounds, 179–180
aversive reaction, 166
conditioned stimuli (CSs), 166
diagnostic criteria, 167
emotional reactions, 173–174
emotional stimuli, 179
general anxiety, 175
high sensitivity, 179
HR and galvanic skin response (GSR), 179–180
interventions and treatment, 180–181
learning theory principles, 166
neuroscience-informed conceptualization, 167
panic disorder (with/without agoraphobia), 175–176
phonophobia, 167–168
psychiatric nomenclature, 168
psychopathology, 168–170
social phobia, 167–168
specific phobia (SP), 175
stress tolerance, 166–167
symptom checklist, MSI, 197–200, 198–199*t*
trauma and stress related disorders, 177
trigger sounds, 179–180
two-factor model, 166–167

N

National Institute of Mental Health (NIMH), 168–169
Negative emotionality, 18–22, 82–83
Neurobiology, SPS
  animal studies, evidence from, 122–125
  blood oxygenation level signals (BOLD signal), 118
  functional magnetic resonance imaging (fMRI), 118
  human studies, evidence from, 119–122
Neurolysin, 126
Neuroticism, 6, 22, 90, 113, 121, 168–169

O

Obsessive-compulsive and related disorders (OCRDs), 172, 177
Obsessive–compulsive disorder (OCD), 177
Orienting sensitivity (OS)/openness, 93
Overarousal, 149–150
Overstimulation, 148–152

P

Parental warmth and acceptance scale (PWAS), 114–115
Parenting, 81–86
Periaqueductal gray (PAG), 126
Peripheral sensitivities, SPS
  glucose metabolism and sensations, 126
  hypothalamus-pituitary-adrenal axis, 125
Personal distress (PD), 190–191
Personality disorders, 91–92
Perspective taking (PT), 190–191
Phonophobia, 167–168
Physical health, 89
Planet Earth
  extreme environments, 188–189
  highly sensitive persons (HSPs), 189–190
    aesthetics and subtleties, 194
    coping strategies, 194
    with positive childhoods, 190
    in space missions and outer-planet settlements, 194–195
  human factor, 188
  space travel, 188–189
    benefits of, 194–195
    challenges of, 189–194
Plasticity genes, 57–58
Positive psychology techniques, 88
Posttraumatic stress disorder (PTSD), 177–178
Principal components analysis, 91–92

R

Radiotracer, 123–124
Relationships, social outcomes, 95–96
Relaxation training, 180
Research Domain Criteria (RDoC) initiative, 62–63, 168–170, 169*f*
Respiratory sinus arrhythmia (RSA), 87
Role-playing responses, 160

S

School-based antibullying intervention, 25–26, 88
Self-care, 150
Self-regulation, 21, 121–122
Sensory processing disorder (SPD), 11–12
Sensory processing sensitivity (SPS), 55–56
  in adulthood, 30–36
  assessment
    approaches to, 137–139
    contradictory traits, 139–140
    differential susceptibility, 140–142
  in childhood and adolescence, 22–30
  Dandelions, Tulips and Orchids, sensitivity levels of, 36–41
  environmental and social stimuli, xiii
  and environmental sensitivity (*see* Environmental sensitivity)
  evolutionary basis for, 109–110
  genes
    ADRA2b polymorphism, 117
    BDNF, 117
    cognitive behavioral therapy (CBT), 118
    dopamine, 114–116
    environmental sensitivity, polygenic score of, 117–118
    epigenetics, 111–112
    genome-wide association study (GWAS), 117–118
    met allele, 117
    nonshared environments, 110–111

serotonin, 112–114
shared environments, 110–111
highly sensitive child (HSC)
parent-report scale, 27, 28*t*
self-report scale, 23–26
highly sensitive child rating system (HSC-RS), 28–30, 29*t*
highly sensitive person-parent-report scale, 23
highly sensitive persons (HSPs) clients, working with
culture, 147–148
ethnicity, 147–148
five goals, 146–147
gender, 147–148
nonsensitive therapist, 143–145
therapist and clients, 145–146
history of, 5–10
implications for research, 60–63
issues
emotional regulation, 152–156
low self-esteem, 148
overstimulation, 148–152
Misophonia, xv
neural basis of
animal studies, evidence from, 122–125
blood oxygenation level signals (BOLD signal), 118
functional magnetic resonance imaging (fMRI), 118
human studies, evidence from, 119–122
peripheral sensitivities
glucose metabolism and sensations, 126
hypothalamus-pituitary-adrenal axis, 125
personal relationships
adjustment, 158–160
research on, 157–158
right start, 160–162
perspective for, xv
phenotypic markers of, 19–22
Planet Earth (*see* Planet Earth)
science of, 4–5
symptoms of, xiii
variation in, 7–9
Serotonin transporter (5-HTT), 112–114, 118, 122–126
Serotonin transporter-linked polymorphic region (5-HTTLPR), short allele, 52–53, 82, 87, 96–97, 112–114, 116–117, 122–123, 126–127
Sexuality, 157
Social adjustment, moderators of, 80–86
Social anxiety
behavioral inhibition, 77
continuity and discontinuity in, 79–80
Social anxiety disorder (SAD), 90–91, 174, 176–177
Social cognition, 113–114
Social outcomes
communication, 96–97
relationships, 95–96
Social phobia, 167–168
Social traits, 113–114
Specific phobia (SP), 174
Stressful life events scale, 114–115
Structured Clinical Interview for DSM-IV Axis II screening questionnaire (SCID-II-SQ), 91–92

T

Teacher-report scale, 30
Temperament, 11, 63, 157
Tinnitus retraining therapy (TRT), 180
Trichotillomania (TM), 177

U

Unconditioned response (UCR), 166
Unconditioned stimulus (UCS), 166

V

Vantage sensitivity (VS), 18–19, 25, 55, 65–67, 88, 95–96
Vasopressin-1a receptor density, 78
Ventral attention network, 193
Ventral tegmental area (VTA), 95
Visual aesthetic sensitivity test, 43
Visual discrimination task, 119–120

W

Work satisfaction, highly sensitive adults, 93–94

FSC
www.fsc.org
MIX
Papier | Fördert gute Waldnutzung
FSC® C083411

Kolibri 360 GmbH
Ferdinand-Jühlke-Straße 7
99095 Erfurt, Deutschland
produktsicherheit@kolibri360.de